Chemistry Research and Applications

nova
science publishers
www.novapublishers.com

Chemistry Research and Applications

Properties and Uses of Antimony
David J. Jenkins (Editor)
2022. ISBN: 979-8-88697-081-4 (Softcover)
2022. ISBN: 979-8-88697-088-3 (eBook)

The Science of Carbamates
Güllü Kaymak (Editor)
2022. ISBN: 978-1-68507-708-2 (Softcover)
2022. ISBN: 978-1-68507-872-0 (eBook)

Deep Eutectic Solvents: Properties, Applications and Toxicity
Carlos Eduardo de Araújo Padilha, PhD, Everaldo Silvino dos Santos, PhD, Francisco Canindé de Sousa Júnior, PhD, Nathália Saraiva Rios, PhD (Editors)
2022. ISBN: 978-1-68507-719-8 (Hardcover)
2022. ISBN: 978-1-68507-799-0 (eBook)

Polycyclic Aromatic Hydrocarbons: Sources, Exposure and Health Effects
Warren L. Gregoire (Editor)
2022. ISBN: 978-1-68507-626-9 (Softcover)
2022. ISBN: 978-1-68507-685-6 (eBook)

Cyanide: Occurrence, Applications and Toxicity
Bill M. Torres (Editor)
2022. ISBN: 978-1-68507-619-1 (Softcover)
2022. ISBN: 978-1-68507-670-2 (eBook)

A Closer Look at Carvacrol
Zak A. Cunningham (Editor)
2022. ISBN: 978-1-68507-627-6 (Softcover)
2022. ISBN: 978-1-68507-634-4 (eBook)

More information about this series can be found at
https://novapublishers.com/product-category/series/chemistry-research-and-applications/

Waldemar Nyström
Editor

The Future of Biorefineries

nova science publishers
www.novapublishers.com

Library of Congress Cataloging-in-Publication Data

ISBN: 979-8-88697-524-6

Published by Nova Science Publishers, Inc. † New York

Contents

Preface

This book is a collection of four chapters which review the future prospects for biorefineries. Chapter One is a review of the future of biorefineries from a circular economic perspective. Chapter Two is a prospective analysis of the challenges which biorefineries in Latin America face. Chapter Three reviews the advances and challenges to the process scale up for biorefineries in Latin America. Chapter Four is a spatially explicit assessment of a forestry biorefinery for the recovery of bio-based materials. The book concludes with a full bibliography.

Chapter 1 - Over the last decade, the world has been facing climate changes because of the environmental impacts caused by excessive use of fossil fuels, a linear economy, and the consequent depletion of natural resources. For these reasons, international organizations established the Sustainable Development Goals 2030 (SDGs), which are guidelines designed to improve the socio-economic and environmental conditions through changes in economic development, to reduce environmental pressure by mitigating greenhouse gas emissions and uncontrolled waste generation. Consequently, the second-generation (2G) biorefinery concept (focused on waste biomass as feedstock) has gained researchers' attention worldwide. Indeed, the potential of underutilized waste as feedstock is an exciting strategy for changing from a traditional economy to a circular one. It enables the diffusion of sustainable development concepts and encourages the creation of new industrial models that consider the social and environmental demands. This chapter discusses the role of biorefineries that uses biomass waste in economics and social progress to meet the SDGs and the circular economy concept. In this context, the authors have discussed SDGs and government policies, biobased products and their importance to market expansions, and the relationship between the advances in industry 5.0 and biorefineries.

Chapter 2 - Biorefineries play an essential role in sustainability and the biobased economy. Some of the challenge related to policy regulation, new process, and raw materials needs structural change in the value chain, such as logistics, discovering new raw materials, and overcoming troubles that arose from pandemic SARS Cov 2. It was evident the need to have an auto sustainable economy, the active participation of quadruple helix efforts to drive new developments and reach new milestones in technological maturity.

Government efforts have facilitated the legal framework for the operation of Research Centers with social projection and have allowed, for example, in Costa Rica, to join the developments of nanotechnology to one of the lines of biorefineries where pineapple waste has been taken advantage of it. On the other hand, advances in this sense in Uruguay and Chile have allowed substantial advances in using residues from the meat industry and the production of polyhydroxyalkanoates, respectively. Undoubtedly, Brazil is experiencing an important boom in the Latin American economy given its long history in the use of agricultural residues and in obtaining biofuels. Colombia is positioned as an important international actor due to its contributions, the adoption of the 2030 bioeconomy agenda, and the governmental impulse through the 2030 bioeconomy mission for Colombia, which requires the cohesion of the actors from the different industrial sectors, academia, the state, and civil society. In this sense, initiatives developed in post-conflict areas in Colombia for the reconstruction of the social fabric and to provide new energy sources and biobased products from organic agricultural and agro-industrial residues in the Colombian Caribbean are also presented. With the aim of performing a prospective analysis of the state of biorefineries in Latin America, this work was divided into the following section.

Chapter 3 - In a Latin American approach (Caribbean, Central and South America), Brazil is a focus of research and development of biorefining processes. Researchers from that country have done remarkable work on first and second generation of biofuels just like Colombia on first generation biofuels. The research has been associated to sugarcane and oil palm transformation and their main residues (e.g., vinasse, bagasse, empty fruit bunches, etc.). Therefore, dealing with lignin and cellulose materials remains the main research challenge. Biomass availability is related to agricultural production. The largest agricultural economies in Latin America after Brazil are Mexico, Argentina, and Colombia. Different biorefinery options have been proposed in these countries to treat agricultural waste. It applies both to extensive crops and to smaller-scale crops and their residues. In this sense, this work addresses some challenges for process scaling up, and the search for sustainable biorefineries in the Latin America context concentrating on the following topics, i) stabilization of biorefinery feed, ii) technoeconomic analysis focused on finding alternatives or technologies for viable projects and iii) analysis of stakeholders in biorefineries. Other important elements discussed are the emerging microorganisms that could play a pivotal role in scaling process and the need to integrate environmental concepts and supply chain in biorefineries. In conclusion, there are opportunities for the development of biorefineries in Latin America taking advantage of the Brazilian experience and based on agricultural practices that may be common in the region. Research and development of new unit "bio" -operations should be strengthened.

Chapter 4 - The biorefinery concept emerged as an alternative to decrease the greenhouse gases emissions by the industrial section while producing bioenergy and value-added products. This chapter evaluated a critical analysis of the residues derived from eucalyptus harvest to support decision-making for a biorefinery implementation in the Administrative Region of Campinas (ARC), São Paulo State, Brazil. Brazilian official data for forestry extraction were collected for the eucalyptus production areas of the municipalities of ARC. RapidEye images were used to process, identify, and classify the areas occupied by eucalyptus and the chemical industries, aiming to compose a georeferenced database. Although most of the developments have been focused on processing technologies, it is critical to define the specific biobased products that should be produced considering market drivers such as demand, production flexibility, logistics, and supply chain configuration. The conversion of lignocellulosic materials present in forestry residues could produce several bio-based products after a pretreatment step. These products could replace fossil-based raw materials for the chemical industry, such as polyurethane, lignin, tannins, furfural, xylose, xylitol, polyphenols, and lubricants. It was possible to observe the existence of promising potential for the use of eucalyptus residues to produce a wide variety of chemicals. In the case of biofuels, the use of residual biomass is a primary factor for the transition to a viable and more renewable energy chain since eucalyptus residues are still little or not explored.

Chapter 1

Future of Biorefineries from a Circular Economy Perspective

Thamarys Scapini[1,2], Natalia Klanovicz[2,3], Aline Frumi Camargo[2,4], Suzana Fátima Bazoti[2,5], Matheus Cavali[1,6], Charline Bonatto[2], Letícia Paliga[2], Andressa Warken[2], Sérgio Luiz Alves Júnior[4,7], Adenise Lorenci Woiciechowski[1] and Helen Treichel[2,*]

[1]Federal University of Paraná, Department of Bioprocess Engineering and Biotechnology, Curitiba, Brazil
[2]Laboratory of Microbiology and Bioprocesses, Federal University of Fronteira Sul, Erechim, Brazil
[3]Research Group in Advanced Oxidation Processes (AdOx), Department of Chemical Engineering, Escola Politécnica, University of São Paulo, São Paulo, Brazil
[4]Federal University of Santa Catarina, Graduate Program in Biotechnology and Biosciences, Florianópolis, Brazil
[5]Federal University of Santa Catarina, Department of Chemical and Food Engineering, Florianópolis, Brazil
[6]Federal University of Santa Catarina, Department of Sanitary and Environmental Engineering, Florianópolis, Brazil
[7]Federal University of Fronteira Sul, Laboratory of Biochemistry and Genetics, Chapecó, Brazil

Abstract

Over the last decade, the world has been facing climate changes because of the environmental impacts caused by excessive use of fossil fuels, a

---

[*] Corresponding Author's Email: helentreichel@gmail.com.

In: The Future of Biorefineries
Editor: Waldemar Nyström
ISBN: 979-8-88697-524-6

linear economy, and the consequent depletion of natural resources. For these reasons, international organizations established the Sustainable Development Goals 2030 (SDGs), which are guidelines designed to improve the socio-economic and environmental conditions through changes in economic development, to reduce environmental pressure by mitigating greenhouse gas emissions and uncontrolled waste generation. Consequently, the second-generation (2G) biorefinery concept (focused on waste biomass as feedstock) has gained researchers' attention worldwide. Indeed, the potential of underutilized waste as feedstock is an exciting strategy for changing from a traditional economy to a circular one. It enables the diffusion of sustainable development concepts and encourages the creation of new industrial models that consider the social and environmental demands. This chapter discusses the role of biorefineries that uses biomass waste in economics and social progress to meet the SDGs and the circular economy concept. In this context, we have discussed SDGs and government policies, biobased products and their importance to market expansions, and the relationship between the advances in industry 5.0 and biorefineries.

Keywords: bioeconomy, green chemistry, 5th industrial revolution, waste resources, bioproducts

1. Introduction

Due to growing environmental concerns related to human pressure on natural resources and population growth, new strategies have been sought to direct efforts to protect the environment and human health. These problems have been the focus of discussions in international organizations, the most prominent attempt being the development of the 2030 Agenda, composed of 17 sustainable development goals (SDGs), which are a series of proposals and targets to reduce the adverse effects of the excessive use of non-renewable resources to improve the quality of life everywhere (ONU 2015; Ubando, Felix, and Chen 2020; Solarte-Toro and Alzate 2021).

Propelled by growth supported by a more conscious production and consumption, discussions about circular economy and bioeconomy have emerged as an alternative to overcome the current demands. Developing cleaner economies that target the entire process beyond obtaining the product is necessary to achieve sustainability. In this scenario, biorefineries based on plant biomass feedstock emerge as an alternative to reduce the pressure on natural resources and dependence on fossil resources. In integrated and optimized processes, biorefineries appear as a strategy to maximize feedstock use and minimize environmental impacts associated with converting high-value-added products (Cherubini 2010). Integrating biorefinery concepts into residue management — the so-called second-generation (2G) biorefinery — is a promising approach for the transition from a linear to a circular economy. While in the former, there is product generation and waste disposal, in the latter, the necessary processing of feedstock is targeted with fewer residues and effluents generation (Duan et al. 2022).

In practical terms, developing an economically viable and sustainable biorefinery with full use of biomass is challenging. However, with the increasing advances in biotechnology, it is possible to integrate biorefining processes (Duan et al. 2022). This scenario is guided by the advancements and transformations caused by Industry 4.0. With new technological revolutions, Industry 5.0 presents itself as a revolution for the future and must take advantage of human creativity to develop more intelligent, accurate, and efficient machines (Maddikunta et al. 2022). With the advancement of technology, a greater focus on productivity has expanded research on the development of bioprocess engineering, resulting in more efficient and resilient production, considering changes in manufacturing processes and the design of machines and facilities, which reduce energy and resource consumption (Xu et al. 2021.

The increase in production efficiency in Industry 5.0 will come with the synergy between humans and machines, which can generate increasingly personalized and sustainable products. The predominant idea is to create green solutions by the industry, which must be a protagonist in preserving the environment. By integrating the biorefinery concept, significant advances may be seen in terms of new operations being added to the existing processes, using the by-product of the industry as raw material for energy generation and new products with higher value-added (Maddikunta et al. 2022). Nanotechnology, bioplastics, organic acids, food compounds, and biofuels are among the products with the most significant potential for integration in biomass biorefineries, which currently presents an excellent tendency for expansion in the economic market increasing the viability of these systems.

From product manufacturing to integrated process management, biorefineries are one of the most promising existing strategies for expanding the economy more sustainably, aiming to achieve the SDGs and ensure the sustainability of the planet. Considering this scenario, this chapter addresses an overview of residual biomass biorefineries, emphasizing future perspectives and market expansion trends of the leading products that can be obtained in these systems. Moreover, we also focus on associations between circular economy, bioeconomics, and biorefineries to achieve the SDGs and a more environmentally sustainable future.

2. From Sustainable Development Goals 2030 to Bioeconomy: The Importance of Biorefineries

The circular economy and the bioeconomy are widely spread concepts in sustainability, which are not yet clearly defined in theory but have been gaining space in scientific and political discussions since they aim at minimizing the environmental impacts resulting from anthropic expansion and population growth. Both are in the direction of a long-term sustainable bio-based economy, culminating in sustainable development, aiming to supply the necessities of the present without compromising the demands of future generations (GBS 2020; D'Amato and Korhonen 2021; United Nations 2015).

In this scenario, the bioeconomy can be described as the production and use of biomass aiming at reuse, recycling, and remanufacturing through new technological processes, generating new products for a sustainable economy (Calicioglu and Bogdanski 2021; Ubando, Felix, and Chen 2020). For Solarte-Toro and Alzate (2021), the bioeconomy is an essential change to achieve sustainability. The circular economy is guided by minimizing the use of non-renewable sources and extracting natural resources as raw materials for processes, providing sustainable alternatives to the current model of economic development (linear economy) (Calicioglu and Bogdanski 2021). For Ubando, Felix, and Chen (2020), the circular economy can be discussed as a closed-loop sustainable framework. The authors highlight biomass as a pillar for the transition from a linear to a circular economy.

Regardless of the theoretical definitions, the principles that guide the bioeconomy and circular economy are transforming the economic, social, and environmental configurations towards systems with fewer losses within supply chains. In this context, the main challenges worldwide are the more efficient and sustainable use of natural resources and reducing air pollution (Nazarko, Chodakowska, and Nazarko 2022). Therefore, there is a strong relationship that integrates these concepts, inferring the macro discussions involving politics, scientific research, and economics that search for different mechanisms to approach the economic, social, and environmental goals, enabling sustainable transformation alternatives (D'Amato and Korhonen 2021).

Given the challenge of promoting sustainable economic growth associated with environmental and human wellbeing concerns, international organizations have proposed several goals and guidelines to reduce the negative impacts resulting from the use of non-renewable resources. Of these efforts, the most prominent is the creation of the 2030 Agenda. It is guided by the 17 SDGs that aim at the collective promotion of development to protect the planet from anthropic degradation and manage consumption and production models in an integrated and equilibrated pathway, thereby creating conditions for long-term sustainable economic growth (GBS 2020; Solarte-Toro and Alzate 2021; Calicioglu and Bogdanski 2021). The 2030 Agenda was signed in 2015 by all UN member states. Adopting these goals encourages economic advances and guides the development of a more sustainable production and consumption scheme (Solarte-Toro and Alzate 2021).

New forms of production and the complete and optimized use of raw materials contribute significantly to achieving the SDGs (Sadhukhan et al. 2018). The success in achieving the goals is combined with preserving natural resources. If the environment is maintained and humans find ways to live in harmony with the use of natural resources, the environmental impact will be minimized.

These objectives are interconnected, and there is a way to conduct them in a form that attends to the necessities of each society because a strategy aimed to reach an objective directly or indirectly will impact others (United Nations 2015; Sadhukhan et al. 2018; Calicioglu and Bogdanski 2021). Most SDGs are directed toward clean production, sustainable agriculture, energy availability, innovative alternatives for business, and conscientious production and consumption. All these concepts are directly related to biorefineries, which can be considered the basis for achieving the SDGs (Conteratto et al. 2021).

The biorefinery concepts distinguish between products, processes, and design, culminating in the valorization of biomass resources, production of economically competitive products, and processes that mitigate environmental impact. As a positive consequence, these processes promote the development of an economy less dependent on non-renewable resources. Associated with the bioeconomy and circular economy concepts, the biorefineries have to include sustainability as an essential part of the project. It must align the strategies to the SDGs, using the feedstock in an optimized pathway, with less generation of effluents and consumption of drinking water, and mainly, making the processes attractive in comparison to techniques with non-renewable resources (Solarte-Toro and Alzate 2021; Conteratto et al. 2021; Ubando, Felix, and Chen 2020).

Biorefineries are a sustainable strategy for producing multiple products from a feedstock, usually associated with vegetal biomass (lignocellulosic) through conversion technologies. They are an important way to make the full implementation of a circular economy (Ubando, Felix, and Chen 2020). The complexity of the integration and holistic use of biomass in biorefineries is still a tremendous challenge for applied science, and several studies are currently searching for high yields and optimal process performance from different biomasses to obtain chemicals and bioenergy (Ubando, Felix, and Chen 2020; Solarte-Toro and Alzate 2021).

Associated with the SDGs, biorefineries can be linked to various strategies to achieve the proposed goals. Several SDG indicators can be impacted by the development of biorefineries, such as poverty eradication (when associated with income generation and jobs) and access to energy. Moreover, it is worth noting that biorefineries may be decentralized, with different scales, depending on the feedstock to be processed and the production capacity to be implemented, which makes them feasible to be installed even in remote areas (Solarte-Toro and Alzate 2021). On a small scale, biorefineries have numerous processing benefits, such as reduced transportation and secondary disposal costs. On the social level, biorefineries contribute to job opportunities, such as in rural areas. Encourage more significant heterogeneity in agricultural production, contributing to biodiversity and highlighting the importance of small farmers in a circular economy. It is important to note that the successful implementation of a local biorefinery will positively impact the macroeconomic indicators (D'Amato and Korhonen 2021; Igbokwe et al. 2022; Bruins and Sanders 2012).

In addition, by promoting conditions for the eradication of poverty and improvements in agricultural situations, it acts directly on the reduction of world hunger, on health and human wellbeing, can be promotion agents for the access to sanitation and drinking water, through projects considering the water cycle, besides facilitating access to energy. In addition to using local biomass, biorefineries reduce capital expenditures by implementing and supplying local market demands through versatile processes (Hassan, Williams, and Jaiswal 2019). By developing an optimized biorefinery for the holistic use of biomass and associated residues, positive results are directed to the SDGs linked to economic growth and more sustainable cities, through a lower dependence on fossil resources, besides resulting in actions to mitigate climate change (Solarte-Toro and Alzate 2021; ONU 2015). In this sense, biorefineries have a crucial role in promoting SDG compliance.

3. Trends and Challenges in Producing Biobased Products from Waste Resources

There is a growing interest in integrating industrial processes for the holistic use of waste biomass. Biorefinery is the central strategy in transitioning from a linear to a circular economy. Although they mainly focus on energy production, biorefining processes based on plant biomass are promising and have expanded to other products of commercial interest (Hassan, Williams, and Jaiswal 2019).

The focus is generally on the study of biological routes for the recovery of bioproducts, mainly aimed at reducing the water and carbon footprints, the use of chemicals, and the generation of effluents, which has been crucial to driving the development of biorefineries. Even if a biorefinery uses residual feedstock to obtain a series of energies and products, it does not mean its process is automatically self-sustainable. To achieve sustainability and minimize environmental impacts, the process can be evaluated holistically and find ways to make it feasible within the existing context. The life cycle analysis (LCA) can determine the environmental impact of biorefinery. This analysis is done by establishing previous limits. Still, generally, the study is done from the feedstock to the product leaving the factory gate, which gives a realistic idea of the product's life cycle (Joglekar et al. 2019; Teh, Tan, and Chew 2021).

Since biorefineries need to be focused on a circular economy, giving support for sustainable economic development. In this context, this topic will address some of the leading products that can be obtained using biomass as feedstock, which presents expansion in scientific research and interest and economic market expansion. Recent information about the market and production of bioplastics, nanomaterials, biofuels, organic acids, and oligosaccharides will be covered in this topic.

3.1. Bioplastics

Bioplastics are polymers produced by biological processes using renewable resources such as biomass, bioderived intermediates, and microorganisms or a combination of these (Nandakumar, Chuah, and Sudesh 2021; Rosenboom, Langer, and Traverso 2022). Many wastes have been studied to produce bioplastics at laboratory and pilot scales. However, commercial-scale production of bioplastics still uses expensive raw materials, such as sugars, whose use and sustainability are widely discussed. It has already been reported that the type of substrate used by microorganisms accounts for about 50% of bioplastics' production cost (Khatami et al. 2021; Lhamo, Behera, and Mahanty 2021).

Biodegradable bioplastics include polylactic acid (PLA), polyhydroxy-alkanoates (PHA), biocellulose, and starch, while non-biodegradable ones include biopolyethylene terephthalate (bio-PET), polyol-polyurethane, and biopolyethylene (Bio-PE) (Narancic et al. 2020). PHA and PLA represented 1.8 percent and 18.9 percent of global bioplastic production capacity in 2021 and are reported to be the main contributors to the growth of biodegradable bio-based plastics (European Bioplastic 2021).

PHAs are bio-based polymers that can be produced by microbial fermentation. They have drawn significant attention as an alternative to petrochemical-based plastics because they exhibit biodegradability, biocompatibility, and comparable physicochemical properties (Khatami et al. 2021; Lhamo, Behera, and Mahanty 2021).

Although higher PHA productivity is obtained when using glucose as a substrate, starch has been used as an alternative substrate since it is present in many food and agricultural wastes and can be an easily accessible and low-cost alternative (Lhamo, Behera, and Mahanty 2021). Starch-based bioplastics accounted for 16.4% of global bioplastic production capacity in 2021 (European Bioplastic 2021). However, to improve plasticity and thermoplastic characteristics, naturally occurring starch is processed by mixing with plasticizers such as glycerol, urea, sorbitol, or glycerin in the presence of high temperatures and shear forces (Khan et al. 2017). Examples of starch-based bioplastics are Bio-PET and thermoplastic starch (TPS) (Nandakumar, Chuah, and Sudesh 2021).

PLA is obtained by polycondensation of lactic acid, which can be derived from the fermentation of sugars present in biobased sources such as corn and sugarcane (Narancic et al. 2020; Ghomi et al. 2021). The starch in corn is also the primary substrate source for PLA production. Currently, almost 51% of PLA production is based on corn as feedstock (Ghomi et al. 2021), although it has been shown the potential of different agricultural biomasses, such as barley, wheat, and brown rice (Oh et al. 2005; Okano et al. 2017), as well as fruits like cashew and red pitaya (Silveira et al. 2012; Yien Ong et al. 2012).

Other substrates have been used for PHA production, such as lignin (present in the structure of lignocellulosic materials), triacylglycerols (present in animal fats and vegetal oils), and lactose (present in whey and microalgae) (Mastropetros et al. 2022; Lhamo, Behera, and Mahanty 2021). Sewage sludge has also been studied for producing PHAs and is economically viable, with a minimum cost of 1.26 and 2.26 US$/kg raw PHA for large and small sewage treatment plants, respectively (Crutchik et al. 2020).

In the face of worsening plastic pollution, bioplastics are increasingly being touted as a green alternative. However, a product should not be considered a green alternative simply because it is obtained from renewable resources. Sustainability depends on how it is got, how it is applied, and how the bioproduct can be recycled. Although bioplastics are derived from renewable resources, this does not indicate that they are biodegradable, i.e., biobased does not mean biodegradable, and some bioplastics are not biodegradable. Those that are biodegradable and compostable require specific conditions for optimum degradation and should not even be discarded with fossil-based plastics. At the same time, non-biodegradable bioplastics can be mixed and recycled with conventional plastics because they are identical polymers in terms of properties. Thus, a critical point is the disposal of these materials in the environment because when done carelessly, it can have the same harmful effect as traditional petrochemical plastics. On a large scale, these attitudes may offset the positive impacts that bioplastics can achieve (Ferreira-Filipe et al. 2021; Nandakumar, Chuah, and Sudesh 2021).

According to life cycle assessment, bioplastics are advantageous in saving fossil resources and contributing to reducing greenhouse gas emissions. However, when the useful life of PLA-based products comes to an end, they are landfilled or composted, and the lack of adequate infrastructure for processing PLA leads to limitations in recycling (Ghomi et al. 2021; Vea et al. 2021).

In addition to environmental sustainability, for a product to be able to dominate the market, financial viability must be achieved. Although wide varieties of bioplastics are available on the market, they have not been able to eliminate fossil-based plastics (Rosenboom, Langer, and Traverso 2022; Nandakumar, Chuah, and Sudesh 2021). One kilogram of polylactic acid (PLA) filaments can cost approximately 16 times more than conventional plastics like polypropylene (PP) and 12 times more than polyurethane (PU) (Nandakumar, Chuah, and Sudesh 2021). Because of this, the global production of fossil-based plastics is still much higher than that of bioplastics: 380 million tons versus 2.2 million tons per year, respectively, demonstrating how far we are still from reaching an ecologically sustainable scenario when it comes to the use of bioplastics as promising substitutes for petrochemical-based plastics (Rosenboom, Langer, and Traverso 2022; BBC Research 2021; Magalhães Júnior et al. 2021).

The global volume of bioplastics is expected to reach 3.3 million tons by 2026 with a compound annual rate of 8.3% (BBC Research 2021), driven by market regulations and increasing consumer demand for sustainable products. Large organizations and companies such as Alliance to End Plastic Waste, Nestlé, Peugeot Citroën SA, and Toyota have committed to developing and producing sustainable plastics. The most considerable growth can be expected for polymers such as bioPE that can be processed on standard equipment and PLA-cellulose blends that are cost competitive on a large scale. However, the demand for PLA- and PHA-based bioplastics currently exceeds supply, as there is a strong demand from the food industry to use these materials in their packaging. Like conventional plastics, the primary market for bioplastics is based on use in single-use items such as food packaging and fast-moving consumer goods with short to medium lifetimes (Rosenboom, Langer, and Traverso 2022). PHAs, for example, is applied in packaging, latex paint production, the agricultural sector, biofuels, and biochemical building blocks (Khatami et al. 2021; Chen 2009; Mohammadi et al. 2012). PLAs are used in packaging, textiles, plasticulture, service tools, and environmental remediation films and have a wide application in the biomedical area (Ghomi et al. 2021; Castro-Aguirre et al. 2016).

Therefore, even as a product with high biotechnological potential, bioplastics mainly produced from lignocellulosic biomass are still limited to applied research, aiming to elucidate mechanisms involved and pilot-scale processes. In this scenario, biorefineries can be a great impulse to this sector since obtaining different products in an integrated approach can enable the production of bioplastics since there is a social tendency to demand cleaner products environmentally, as discussed in the topic of this chapter on the basis that guides the industry 5.0th.

Considering this, the bioplastics market has been showing growth, and the demand for new bioplastics is expected to be driven by regulatory incentives and circularity-oriented policies. However, in addition to the production and insertion of bioplastics, one should vigorously enforce reuse and recycling routes of these new bioproducts to extend their lifetime and ensure a circular and low carbon economy (Rosenboom, Langer, and Traverso 2022).

3.2. Biofuels

Usually, when discussing biorefineries, one refers to the production of biofuels since they are the primary and most recurrent products derived from biomass in cascade processes. Biofuels can bring energy autonomy to biorefineries, besides being high-value-added products with an established economic market. Biofuels are essential to developing a circular economy based on the biorefinery concept. This section will address five known biofuels: ethanol, biodiesel, biogas, biohydrogen, and butanol, from a biorefinery perspective.

3.3. Ethanol

In 2020, the global ethanol market was valued at USD 93.7 billion, predicted to exceed around USD 155.6 billion by 2030. Among the main factors responsible for that growth is the increased utilization of ethanol as biofuel (Research 2022). The three countries leading ethanol production are the United States (48.2%), Brazil (26.2%), and China (8.1%). By 2029, the global ethanol production is projected to increase to 140 billion liters, and sugarcane and maize will continually dominate as the primary feedstocks: ethanol generation is predicted to utilize 25% and 14% of the total global sugarcane and maize production, respectively. Currently, ethanol production is based on maize (64%), sugarcane (26%), molasses (3%), and wheat (3%), while the remaining comprises other grains, cassava, or sugar beets (Food and Agriculture Organization 2020). The two primary feedstocks utilized for ethanol production are exploited owing to their readily available carbohydrates: corn grains and sugarcane juice. But biorefinery approaches are still necessary for the first-generation (1G) ethanol industry, and they could be implemented in the processes currently developed.

The co-production of marketable products beyond ethanol can be highly beneficial for the ethanol industry if the feedstock comes from its byproducts. For instance, Gerrior et al. (2022) evaluated the integration of corn-ethanol production with anaerobic digestion of the thin stillage to divert this byproduct from energy-intensive evaporators, followed by struvite recovery. The resulting digestate was then aerobic treated and proposed for microalgae cultivation. The process water containing algae was recycled as a source of starch and nutrient for the corn-ethanol plant. This closed-loop system reduced energy consumption by providing methane and avoiding natural gas usage, in addition to increasing co-product generation, such as struvite, that can potentially replace fertilizers (Gerrior et al. 2022). Other studies have also assessed different approaches to the corn-ethanol industry, such as the simultaneous saccharification and fermentation of microalgae and corn starch for ethanol and biopeptides production (Astolfi et al. 2020). Moreover, the coupling of production technologies for sustainable aviation fuel with dry grind corn ethanol mill has also been analyzed (Tanzil et al. 2021).

Concerning the sugarcane-ethanol industry, integration approaches have also been studied to develop the biorefinery concept in these facilities. The ethanol production from sugarcane generates different residues: sugarcane straw after harvesting, ashes from bagasse combustion in cogeneration plants, filter cake from juice clarification, vinasse after ethanol distillation, and CO_2 from bagasse combustion and ethanol fermentation. All these byproducts are potential feedstocks for producing value-added products. Following a cascading approach, the organic residues can either be converted into bioplastic, carboxylates, and bio-fertilizer or fed anaerobic digesters for biofuel production. The biogenic CO_2 could be commercialized for other industries, such as chemical, pharmaceutical, and cosmetics. Besides, even the silica from sugarcane residues and bagasse ashes has the potential for profitable applications (Formann et al. 2020). Furthermore, the sugarcane industry might be coupled with thermochemical processes to produce liquid, solid, and gas fuels. When the sugarcane-ethanol sector is combined with hydrothermal liquefaction (Deuber et al. 2021), hydrothermal carbonization (Santos et al. 2021), or pyrolysis (Miranda et al. 2021) are possible alternatives for biorefinery development in this sector.

In addition to corn grain and sugarcane juice, ethanol production in a biorefinery approach can also be performed with different feedstocks. All those common feedstocks used in the 1G ethanol industry present a lignocellulosic structure (cellulose, hemicellulose, and lignin) that remains as residues in the fields or industry after processing (Sherpa et al. 2022; Sharma, Jha, and Panwar 2021; Bohn et al. 2021). Thus, the conversion of those residues into 2G ethanol and value-added products in biorefinery plants could be a profitable activity.

For instance, the simultaneous production of ethanol and adipic acid (a high-value biochemical required worldwide) from corn stover was economically and environmentally evaluated. The study obtained ethanol from cellulose and adipic acid from hemicellulose, revealing a suitable biorefinery process by coproducing a biofuel and a value-added chemical (Choe et al. 2021). Furfural was another molecule of commercial interest produced beyond ethanol from corn cobs in an integrated biorefinery approach. A treatment with formic acid yielded a cellulose-rich fraction, which was post-treated, enzymatic hydrolyzed, and fermented into ethanol. The formic acid in the spent liquor was then distilled, and lignin was recovered, providing a hemicellulosic syrup for furfural production (Qiao et al. 2022). In a pilot-scale biorefinery, another study assessed ethanol production from paddy residues (rice straw and rice husk) while recovering silica, lignin, and nutrients (Le et al. 2022). It could raise the ethanol-industry income given the potential applications of both silica (Nayl et al. 2022) and lignin (Yu and Kim 2020; Cavali et al. 2021). This would then promote the economic sustainability of the lignocellulosic ethanol industry by providing other commercial products and assisting biorefinery development. In the cellulosic-ethanol sector, for example, the pretreatment step generates a hemicellulosic hydrolysate that could be microbially converted into other products in addition to ethanol, such as acetic acid, xylitol, single-cell protein, lipids, and carotenoids, reducing waste generation and improving the revenues (Soltaninejad, Jazini, and Karimi 2022). Therefore, every stream of the ethanol production process from different feedstocks can be valued through biorefinery procedures. This route might bring not only economic gains but also environmental benefits.

The LCA of ethanol biorefinery has presented good results. The reduction in energy utilization and the increased co-product production mitigated the environmental impacts in the biorefinery system evaluated by Gerrior and co-workers in the corn-ethanol industry. The novel configuration reduced the environmental effects in the following categories: global warming, land use, fossil resource use, acidification, and water use (Gerrior et al. 2022). Another study considering liquid hot water pretreatment of switchgrass also showed that a biorefinery scenario (production of ethanol, electricity, furfural, acetic acid, and formic acid) was better regarding environmental impacts than producing only ethanol and electricity by attending to greenhouse gas reduction requirements (Larnaudie, Ferrari, and Lareo 2021). Thus, it implies that biorefinery can soon be an environmental-friend approach to 1G and 2G ethanol industries. However, more research is imperative to evaluate different biorefinery scenarios.

3.4. Biodiesel

The biodiesel market was valued at around USD 46.79 billion at the end of 2021, and it is projected to attain almost USD 51.48 billion by 2026 (Market

Data Forecast 2022). This scenario is driven by global biodiesel production, which is forecasted to reach 46 billion liters by 2029. In this order, the European Union, United States, Indonesia, Brazil, and Argentina are the five major biodiesel producers. And the most common feedstocks are rapeseed oil (37%), soybean oil (27%), used cooking oil (23%), and palm oil (9%) (Food and Agriculture Organization 2020). Commonly, the transesterification reaction is the route performed to convert those feedstocks into biodiesel (Manaf et al. 2019).

From a biorefinery perspective, the current biodiesel industry has the potential to supply other products beyond the target biofuel. Considering the biodiesel from rapeseed oil, the waste glycerol was evaluated to produce polyhydroxybutyrate, a biodegradable plastic with numerous industrial applications (Nitkiewicz et al. 2020). Besides glycerol, rapeseed meal is another byproduct from the biodiesel industry that might undergo biorefining. The proteins and phenolics from the rapeseed meal are suitable for several applications in high-value markets, such as food, cosmetics, and nutraceuticals. This biorefinery approach could divert rapeseed meals from traditional utilization as animal feed by integrating the biodiesel industry with more lucrative sectors (Di Lena et al. 2021).

Biorefinery opportunities for biodiesel industries are also based on soybean oil, used cooking oil, and palm oil. The upstream processes before soybean oil conversion into biodiesel can generate different byproducts, such as soybean hulls and phospholipids (Barreiros et al. 2020; Steinke et al. 2021). Soybean hull, for example, has the potential as feedstock to produce bioethanol, butanol, enzymes, plant growth hormones, prebiotics, and other molecules (Amaro Bittencourt et al. 2021). Phospholipids have applications in food technology, pharmaceutical, signal transduction, and drug delivery systems (Singh, Gangadharappa, and Mruthunjaya 2017). Concerning the valorization of glycerol from biodiesel production with used cooking oil, this byproduct was proposed as the substrate of a microbial consortium to produce 1,3-propanediol and lactate by fermentation (Wang et al. 2019), tested for hydrogen production in batch anaerobic reactors (Rodrigues et al. 2019), and evaluated as growth media for microalgae cultivation (Paladino and Neviani 2022). Likewise, palm oil biodiesel is also related to the generation of different residues: oil palm empty fruit bunches (solid) and palm oil mill effluent (liquid). Those wastes can be utilized to produce other valuable bioproducts, such as biofuels, enzymes, bioactive nanoparticles, and beta-glucans (Letti et al. 2021). Therefore, considering the scenario depicted above, the present biodiesel industry has the possibility of upgrading its upstream and downstream processes into biorefinery approaches, albeit more research is required to validate the economic and environmental gains.

However, apart from that traditional biomass, biodiesel might be obtained from alternative raw materials, such as algae (Bošnjaković and Sinaga 2020; Malik et al. 2022), agroindustrial residual oils and fats (Cavali et al. 2020; Mibielli et al. 2020; Toldra-Reig, Mora, and Toldra 2020), and sewage sludge (di Bitonto et al. 2020; Liu et al. 2021). Those biodiesel feedstocks could raise several opportunities for biorefinery development in this sector. For example, Malik et al. (2022) suggested a cascade biorefining by cultivating microalgae in urban wastewater. In addition to fixing CO_2, the microalgae growth treated the wastewater by reducing its contents of nitrogen, phosphorus, and chemical and biochemical oxygen demands. After harvesting, the microalgae yielded carotenoids and lipids that were transesterified into good-quality biodiesel. The residual microalgae biomass was converted into α-amylase and mycoproteins by fungal fermentation. Accordingly, this biorefinery process resulted in lower waste generation by combining wastewater treatment, biodiesel synthesis, and the production of high-value marketable products (Malik et al. 2022). But for further development of such an approach some optimizations regarding the crucial parameters in this system: oil content of the algae, nitrogen content of the wastewater, and extraction and transesterification efficiencies (Arutselvan et al. 2022).

Concerning biodiesel from residual agroindustrial oils, the fatty acid from the physical refining of soybean oil was proposed as biodiesel feedstock in an enzymatic-catalyzed process. Thus, biodiesel production could be attached to soybean-refining industries to add value to that byproduct (Mibielli et al. 2020). A similar approach was suggested to use poultry wastes for biodiesel production: chicken oil as raw material and eggshell as the catalyst of the transesterification reaction (Odetoye, Agu, and Ajala 2021). The utilization of residual chicken oil (Cavali et al. 2020) and residual bovine fat (Silva and Alhadeff 2022) from meat processing industries are alternative biomass for biodiesel production. Meanwhile, following the biorefinery concept, the glycerol from those processes can be subjected to thermochemical or biochemical routes to obtain other marketable products (Schultz et al. 2014).

Other sources of raw material for biodiesel production are municipal wastewater treatment plants. The sludge and scum from those facilities contain lipids that could be extracted by appropriate solvents and converted into biodiesel (Liu et al. 2021). As exemplified by other studies, it was possible to obtain not only biodiesel but other products from sewage sludge, such as amino acids and protein, short-chain fatty acids, enzymes, bio-pesticides, bioplastics, bio-flocculants, bio-surfactants, and bio-lubricants (W. Zhang et al. 2018; di Bitonto et al. 2020). Besides diverting residues from landfills, using the byproducts from wastewater treatment plants to produce biodiesel might be a step toward biorefinery development in this sector.

Considering the above, the biodiesel industries can present significant potential to operate as biorefineries. The combination of biodiesel production with other high-value products following the biorefinery concept could provide the economic feasibility of this approach (Chia et al. 2018). But to evaluate these biorefineries in the mid-term and long-term, it is advisable to perform life cycle and techno-economic assessments to select the most suitable conditions for each scenario proposed.

3.5. Biogas

In 2018, 59.3 billion m^3 of biogas was produced globally. By region, Europe is the leading biogas producer (52.03%), followed by Asia (32.50%), Americas (14.04%), Oceania (1.41%), and Africa (0.02%) (World Bioenergy Association 2020). Europe's position in the biogas scenario is due to its favorable regulatory policies and government investments. From an economic perspective, the global biogas market is predicted to reach USD 81.37 billion by 2028 (Research and Markets 2021).

Chemically, biogas is a mixture of methane (CH_4), carbon dioxide (CO_2), and small quantities of other gases obtained by anaerobic digestion (AD) of organic matter. And the feedstocks utilized in the AD comprise crops (energy crops, crop residues, and sequential crops), animal manure, municipal solid waste, and municipal wastewater (International Energy Agency 2020). A set of four reactions governs the AD process: i) hydrolysis, which breaks down carbohydrates, lipids, and proteins into sugars, fatty acids, and amino acids, respectively, by extracellular enzymes from fermentative bacteria; ii) acidogenesis, the stage in which soluble compounds are absorbed by acidogenic bacteria and converted into simple compounds, such as volatile fatty acids (VFAs), alcohols, CO_2, H_2, NH_3, H_2S, and new cellular material; iii) acetogenesis, comprising the conversion of the previous products into acetic acid, CO_2, H_2, and also into new cellular material; and iv) methanogenesis (acetoclastic and hydrogenotrophic), performed by methanogenic archaea to produce CH_4 (Merzari et al. 2019; Meegoda et al. 2018). The byproduct of anaerobic digestion is digestate, a liquid or solid material that remains after biogas production (Guilayn et al. 2020).

Accordingly, biogas plants could work as biorefineries. The biogas can be used for domestic cooking, converted into electricity or heat by combined heat and power plants, or upgraded to biomethane (Jain et al. 2019; Primmer 2021). And in addition to energy applications, biogas could alternatively serve as feedstock for polyhydroxybutyrate production. Polyhydroxybutyrate is a biodegradable plastic that can be biologically synthesized from CO_2 or CH_4 (Battista et al. 2020). But beyond the conventional AD product (biogas), other intermediate molecules of commercial interest, such as carboxylic acids (i.e., butyric acid, propionic acid, and caproic acid), might be obtained by the biorefining of the AD process (Eng Sánchez et al. 2021).

Regarding digestate, after suitable treatment, it can be disposed of in the soil for fertilization purposes, given its organic carbon content and nutrients (Jain et al. 2019; Primmer 2021). But apart from the agricultural valorization of digestate (i.e., NPK fertilizers and soil improvers and organo-mineral fertilizers), the AD byproduct could be used for producing biofuels from fermentation processes (bioethanol and biohydrogen), harvested biomass (microalgae cultivation to produce biodiesel), or thermal conversion (pyrolysis, gasification, and hydrothermal carbonization to generate biochar, hydrochar, bio-oil, and syngas) (Guilayn et al. 2020). Concerning those thermochemical processes, AD might be integrated with pyrolysis, gasification, and hydrothermal carbonization to treat the digestate in a biorefinery approach whose biochar and hydrochar can be applied either as additives to increase the AD performance or as adsorbents for biogas purification (Cavali et al. 2022). Furthermore, digestate can be the precursor for many other products, such as bio-pesticides, bio-surfactants, functional nanomaterials, and enzymes (Selvaraj et al. 2020).

Therefore, biogas production has excellent potential for biorefinery development. The scenario is favorable to both mid-term and long-term biorefinery approaches, where the integration of processes will play a vital role in this achievement. From now on, it is crucial to evaluate those possibilities from technical, environmental, social, and economic perspectives to guarantee the feasibility of this biorefinery model.

3.6. Biohydrogen

Biohydrogen is one of the renewable fuels available to supply energy in an environmental-friendly way due to its low environmental impact (zero emission of greenhouse gases) and high calorific value (Younas et al. 2022;

Cheng et al. 2022). Most of the biohydrogen used worldwide is for petroleum refining and industrial uses in chemical processing. The main production route is thermochemistry, with 80% of the hydrogen produced from methane reforming, whose efficiency ranges between 74% and 85%, and it is responsible for about 900 Mt of CO_2 emissions per year (Younas et al. 2022). Gasification and pyrolysis of coal are other ways to obtain hydrogen; however, those processes disadvantageously utilize non-renewable sources, in addition to demanding large amounts of energy and being highly harmful to the environment (El-Qelish et al. 2022; Cheng et al. 2022; International Energy Agency 2022; Pingkuo and Xue 2022).

By 2050, hydrogen is estimated to be responsible for 18% of global energy consumption. Obtaining biohydrogen from biomass or renewable raw materials has drawn attention due to its ecological benefits and energy efficiency. Biochemical processes, such as water electrolysis, pyrolysis of organic waste, and supercritical water gasification, are viable alternatives for producing green hydrogen – biohydrogen (El-Qelish et al. 2022).

Biological processes stand out given their environmental and economic advantages. Biohydrogen production results from microorganisms' activities, mainly bacteria that degrade carbohydrate-based substrates from different residues. By utilizing biomass, biohydrogen can be obtained by several processes, such as bio-assisted photolysis of water by algae and cyanobacteria, photodecomposition of organic compounds by photosynthetic microorganisms, fermentation of organic residues, or even the combination of more than one process (Praptyana and Budiyono 2022; El-Qelish et al. 2022). Biohydrogen production by the biological route is a profitable and promising strategy that boosts the development of a circular and sustainable economy (Iqbal et al. 2022; Banu et al. 2020; Cheng et al. 2022).

For producing biohydrogen, several types of raw materials can be used: forestry, agricultural, animal, and food residues, sludge, industrial effluents, urban solid wastes, and wastewater (Saadatinavaz, Karimi, and Denayer 2021). The production processes are divided into technologies that depend on light (i.e., biophotolysis and photofermentation) and technologies that are independent of light (i.e., dark fermentation and microbial electrolysis) (Dzulkarnain et al. 2022; Cheng et al. 2022; Praptyana and Budiyono 2022).

Regarded dark fermentation, several methods have been studied, such as the improvement of fermenting microorganisms through metabolic changes by genetic engineering, light transfer efficiency, enzymatic activity, and ATP content, the optimization of cultivation conditions, the tolerance of bacteria, the immobilization of microorganisms, the utilization of nanotechnology, use of artificial microbial consortia containing selected microorganisms with specific metabolic functions, among others (Cheng et al. 2022; Dzulkarnain et al. 2022). Additionally, there are new technologies employed in the processes, such as reactor improvement, two-stage process integration, use of membranes, a consortium of microorganisms, and pretreatment of raw material (Zhu et al. 2022; Preethi et al. 2022; Iqbal et al. 2022; El-Qelish et al. 2022).

Biohydrogen production can also be performed in a biorefinery context. An example is the sugarcane processing industry, where it is possible to use the bagasse to produce nanocellulose, lignin, and biohydrogen after processing the sugar. Interestingly, when the stages of the generation processes of those products were compared separately, they all had a more significant impact than in the integrated form in the biorefinery system (Katakojwala and Venkata Mohan 2022). In this context, biohydrogen production in an integrated approach is advantageous since it strengthens the principles of green chemistry and has economic advantages.

Following the biorefinery concept, the production of biohydrogen is an effective strategy. Beyond producing hydrogen, it is possible to recover a wide range of products with high added value while reducing the generation of effluents. Moreover, there exists the possibility of simultaneously obtaining another biofuel from the same route, such as ethanol, which is also produced by fermentation (Banu et al. 2021; 2020), and biodiesel, which could be obtained by integrating biohydrogen production with a chemical process using the same raw material (Panagiotopoulos et al. 2013).

Among the challenges of biohydrogen production are the integration of processes, the heterogeneity of the effluents utilized as feedstocks, and their interactions with the microorganisms used in the reaction. In addition, it is necessary to increase production efficiency to meet commercial demands. Therefore, new technologies, such as membrane-based production, fermentative membrane bioreactors, microbial electrolysis cells, and processes with downstream membranes, have already been studied. Metabolic engineering can also contribute to changes in microorganisms to improve their resistance to pH and temperature (El-Qelish et al. 2022; Banu et al. 2020).

Including biohydrogen production in systems already in operation aiming at utilizing industry byproducts as feedstock within an integrated biorefinery concept is an attractive path to overcome challenges and promote sustainable development (Banu et al. 2020).

3.7. Biobutanol

Butanol is a biofuel with a high capacity to replace fossil fuels since it has lower nitrogen oxides and smoke emissions can be produced from renewable sources, and is environmentally clean. The global production capacity was estimated at 2.8 million tons annually, most of which comes from the petrochemical route (Cai et al. 2022). Compared to other alcohols, butanol is more resistant to friction, has less corrosiveness and flammability, has a lower enthalpy of vaporization and volatility due to high carbon content, presents higher thermal efficiency, and is safe to transport (Gwalwanshi, Kumar, and Chauhan 2022; Liu et al. 2021). Butanol has characteristics like gasoline, making it a promising fuel for replacement without requiring significant engine changes. In addition, it is widely used as a solvent and chemical stabilizer in several areas of industry, and it is a precursor in chemical synthesis processes (Meramo-Hurtado et al. 2021; Liu et al. 2021).

The biological route for butanol production, through acetone-butanol-ethanol (ABE) fermentation, emerged in World War I and was the second largest fermentation industry until the 1980s. Since then, petrochemical routes dominate butanol production, and its obtainment occurs via non-renewable raw material, although it is possible to obtain butanol through biotechnological routes (Cai et al. 2022). As a 2G biofuel, butanol can be produced by fermenting different waste biomass, not competing with human food, and reducing dependence on 1G biofuels (Gwalwanshi, Kumar, and Chauhan 2022).

In a biorefinery, butanol production can co-occur. An example is ABE fermentation, a biorefining method of butanol from lignocellulosic biomass (Liu et al. 2021). In that process, the carbon source can be a wide range of raw materials (i.e., residues rich in glucose, effluents, food scraps, starch-rich biomass, wheat straw, rice, and corn). The reaction occurs by inoculating fermenting microorganisms under mild reaction conditions at relatively low temperatures (Liu et al. 2021; Zhang et al. 2022; Xia et al. 2019; Liu et al. 2021). ABE fermentation can be integrated and optimized in several other routes to achieve higher productivity and less interference of inhibitors generated during the processes. It facilitates the separation of compounds produced in fermentation (Liu et al. 2021). In addition, mixed cultures can be employed in consolidated bioprocesses, co-producing butanol, ethanol, acetone, biogas, and biohydrogen (Sanchez et al. 2017).

Integrating a biorefinery approach to generate other target products, using the effluent for producing butanol, can be a sustainable and promising alternative. However, there still are some challenges to ABE fermentation, such as biomass requires pretreatment to release the sugars from the lignocellulosic matrix, inhibition by reaction products, and the construction of bioreactors (Zhang et al. 2022; Meramo-Hurtado et al. 2021). Additionally, it is necessary to overcome the challenges concerning high operating costs to achieve a sustainable project from economic, technical, and environmental perspectives (Meramo-Hurtado et al. 2021).

Residual biomass from agricultural activities is available worldwide, and the choice of raw material is crucial for biorefinery success (Meramo-Hurtado et al. 2021). Accordingly, biobutanol production by ABE fermentation has been extensively studied as an alternative to producing biofuels from solid waste management. It is crucial to evaluate and compare systems based on biomass processing to enable biobutanol production on an industrial scale.

LCA, socio-technical assessment, and exergetic efficiency analysis are tools to evaluate the feasibility of the processes to implement integrated production for ABE. Regarding biobutanol biorefineries, it is necessary to assess the economics of the process considering different biomasses. Recently, an LCA assessment was evaluated for a biobutanol production biorefinery using palm residue as feedstock. A lower production cost was found by linking an ammonia pretreatment process with a production cost of approximately 1.58 $/L (Mahmud and Rosentrater 2019). In another scenario, cassava residues were evaluated for a biorefinery to produce industrial grade butanol as the main product and to generate analytical grade ethanol and acetone (ABE fermentation). The evaluated stages were pre-treatment, hydrolysis, fermentation, and separation of products. A process with a yield of 0.117 kg of butanol/kg of biomass was modeled, with exciting prospects for development at scale (Meramo-Hurtado et al. 2021). The evaluation of an integrated biorefinery for biofuel production allows obtaining relevant information on chemical processes and finding critical points (i.e., energy consumption, waste produced, cost-effectiveness) to propose changes (Meramo-Hurtado et al. 2021).

Butanol is an alternative to fossil-based fuels due to its characteristics and the availability of multiple potential feedstocks for its production. As a result, this biofuel is promising and presents numerous advantages if performed following the biorefinery concept, being co-generated with other products of high added value. Furthermore, this approach contributes to achieving the SDGs within the context of the bioeconomy, boosting the circular economy and sustainable development.

3.8. Nanomaterials

Nanomaterials are materials with dimensions in the range of 1 to 100 nm. They result from an emerging technology that aims to offer products with more excellent performance and flexibility. Today nanomaterials can be found in coatings, electronics (sensors and equipment), cosmetics, sports equipment, environmental remediation, chemistry, parts, and mechanical equipment, but mainly in industrial biotechnology (Reshmy et al. 2022). The demand for this material is expanding due to its vast application possibilities. According to studies carried out by Global Market Insights Inc. (GMI) regarding the size of the nanomaterials market, there is a trend of development and expansion until at least the year 2028, mainly in the regions of North America and Europe, which may be directly linked to the extensive use of medical imaging techniques to diagnose patients, along with investments in pharmaceuticals and the electronics sector. Nanomaterials can move the global market, as companies such as Altair Nanotechnologies, Southern Clay Products, Fuso Chemicals, DuPont, Bayer, and BASF command the sector and follow clear innovation and technology goals (GMI 2022). An example is that in 2021, Intel revealed plans to invest US$ 20 million to establish two chip factories in Arizona, USA.

The feasibility of obtaining nanomaterials depends on factors such as the availability of feedstock, the possibility of scaling up, production cost, and market analysis. Some studies carried out a technical-economic evaluation of lignin nanoparticles and demonstrated that selling this nanomaterial with a 40% profit/t produced is possible. Each nanomaterial will present different physicochemical characteristics, dependent on the biomass source used when obtained in integrated biorefineries. Generally, nanomaterials have a larger surface contact area than their volume, allowing larger contact areas to create stable and well-performing processes (Baig, Kammakakam, and Falath 2021).

Inside the concept of biorefineries, nanomaterials can be obtained from residual biomasses, mainly composed of cellulose, hemicellulose, and lignin, known as lignocellulosic biomass. Agricultural, forestry, and industrial residues are some examples of potential sources for obtaining nanomaterials. Only 40% of the lignin is used to supply the energy demand in biorefineries, and the remaining 60% has the potential to produce bio-based materials. An infinity of nanomaterials can be obtained from lignin and cellulose, such as nanoparticles, nanofibrils, and nanosheets. These nanomaterials may have essential properties, for instance, in the pharmaceutical sector, acting as antimicrobial and antioxidant. In addition, lignin has been considered in the development of new materials due to its sustainable bias with low price, biodegradability, and low toxicity (Zhu et al. 2021).

Lignin nanoparticles are consolidating due to their versatility. Recently, it has been possible to implement its use in thermoplastics, nanocomposites, foams, bactericides, enzyme immobilizers, drug encapsulators, food packaging, and others (Österberg et al. 2020; Schneider, Dillon, and Camassola 2021).

Just as there are nanomaterials from lignin, there are also those from cellulose. Nanocellulose has relevant characteristics such as rigidity, resistance, hydrophobicity, accepting chemical modifications, biocompatibility, low density, and coefficient of thermal expansion. Due to their numerous significant properties, cellulose nanomaterials can be applied in composites, packaging, coatings, cosmetics, and medical and electronic applications, among others. The nanocellulose market has a lot of potential for expansion, with an annual growth rate of 21.4% from 2020 to 2026, according to Global Market Insights Inc, equivalent to $418.2 million (Squinca et al. 2022).

In the context of biorefineries, the most desirable route to obtain nanomaterials is integrated with bioenergy production and other biomaterials. The transformation of nanomaterials from lignocellulosic sources via enzymatic hydrolysis is advantageous in the non-production of toxic residues, reduction in drinking water consumption, and the possibility of reusing the sugars released during the hydrolysis process to obtain other bioproducts. By enzymatic hydrolysis, nanomaterials produce products with high added value, are functional, and have good stability (Squinca et al. 2020; de Aguiar et al. 2020).

The environmental impact caused by the production of nanomaterials can be evaluated by the LCA, as demonstrated in a recent study where an LCA investigation was carried out in a lignocellulose biorefinery (Koch et al. 2020). In this case, lignin must be extracted and precipitated as nanoparticles. The simulations were carried out for 1 kg of lignin nanoparticles suspended in water. The precipitation and purification parameters have been optimized and varied within the LCA. The other parameters studied were considered constant, considering inputs and outputs of consumption materials and energy. The environmental impacts considered relevant were: the origin of inputs and the supply of thermal energy, ethanol emission in the liquid waste stream, water use, and potential for human toxicity. The indicators of global warming potential and human toxicity were identified as hotspots of the process, with a joint contribution of 80-90% of the impact, while in the purification step, the water input indicator was the most expressive, with an influence of 70-97% under the effect of the process (Koch et al. 2020).

To visualize the flexibility of integrated biorefineries, in another study, bunches of empty palm fruits were used as raw material to obtain a series of products, cellulose nanocrystals, cellulose nanocrystals containing lignin, nanolignin, syngas, electricity, methanol, and ammonia. LCA was applied in the process, and the results were for a functional unit of 1 ton of empty palm fruit bunches. The method of obtaining the nanomaterials was compared, and the cellulose nanocrystals had the most significant environmental impacts in terms of global warming potential, human toxicity, and water consumption. Cellulose nanocrystals containing the lignin process require the use of chemicals. They, therefore, represent the possibility of more significant damage to human toxicity, which leads to a higher final cost of the product and a more substantial amount of water to remove excess chemicals. Nanolignin arises to combat the challenges of producing cellulose nanocrystals containing lignin. As a result, its environmental impact, according to LCA, is the lowest among the three nanomaterials, and water consumption has been significantly reduced. On the other hand, to obtain nanolignin, large amounts of thermal energy are needed, and even so, it becomes more profitable. When nanomaterials are integrated with other products, thinking within a biorefinery context, there is a lower environmental impact and a higher profit margin (Teh, Tan, and Chew 2021).

3.9. Organic Acids

Organic acids (i.e., acetic, citric, formic, lactic, propionic, fumaric, malic, etc.) have increasingly attracted the interest of researchers and investors. Biobased routes have been widely studied and produced these compounds, and growing environmental concerns have boosted this scenario (Fact.MR 2021). The prospect that organic acids production is an exciting strategy in biorefineries is mainly due to the expansion of the global market that was valued at US$ 8.27 billion in the year 2016 and is expected to reach US$ 11.39 billion in 2022, considering acetic, citric, formic, lactic, propionic, ascorbic, gluconic and fumaric acids. This market expansion forecast is mainly associated with applying organic acids in food and beverages, personal care products, and pharmaceuticals (Markerts and Markets 2017).

Among the twelve sugar-based chemical building blocks of chemical or biological conversion pathways of most significant relevance as chemicals identified by the U.S. Department of Energy, nine are organic acids (Werpy and Petersen 2004). This is mainly due to the versatility of these molecules, which can be applied in many different areas and replace petroleum-based products. Even so, at present, the synthetic routes using petroleum as a feedstock are the most used for obtaining some organic acids, such as acetic, succinic, and fumaric. However, with the growing discussion of environmental issues and the development of plant biomass biorefineries, there is an increased interest in biological routes through synthesizing these compounds (e.g., succinic acid). In some cases, bioprocessing presents economic advantages over chemical syntheses, such as for citric, fumaric, itaconic, and gluconic acids (Magalhães Júnior et al. 2021; Dezam et al. 2017).

In the context of the biological route, organic acids are small molecules excreted by microorganisms in synthesis processes. Some organic acids are produced in large-scale biological pathways from first-generation feedstocks, i.e., from culture media rich in sugars (C6 and C5) that microorganisms use to synthesize the compounds of interest, using sugar cane and sugar beets or starchy feedstock. These processes are already scaled up in pilots and industry, such as for citric acid predominantly produced with the fungus *Aspergillus niger* (Ciriminna et al. 2017; Reena et al. 2022).

Second-generation biomasses, usually agricultural residues (e.g., hydrolysates of bagasse, straw, husks, etc.), are the focus of most organic acid production research. However, they are still under development for cost reduction, mainly for pretreatment, aiming to maximize the cleavage of hemicellulose and cellulose from the lignocellulosic matrix, generating sugars such as xylose and glucose, and increasing the process yield (Daful and Görgens 2017; Magalhães Júnior et al. 2021; Kuenz and Krull 2018; Bukhari et al. 2020; Upton, McQueen-Mason, and Wood 2020).

In a study developed by Magalhães Júnior et al. (2021) on the organic acids production by biological route for application in bioplastics, the lactic and gluconic acids were highlighted as the most promising routes using lignocellulosic waste hydrolysate as feedstock, with yields higher than 87%. There is still a long way to explore industrial biomass with the potential for producing organic acids and other products in a biorefinery concept. Although many of these residues are used for heat generation in industries, biotransformation is a more exciting alternative to the surplus and less harmful to the environment.

Advances are expected in the coming years in integrating organic acid production into other products and energies. Among these advances in obtaining acids through biological routes, metabolic engineering is one of the most targeted strategies to increase efficiency. The deletion or overexpression of genes in microorganisms to increase yield and reduce metabolic pathways that may interfere with the results of organic acids production are often discussed (Hou et al. 2018; Lu et al. 2022; Morales et al. 2016).

The techno-economic analysis of organic acid production has been a challenge since it needs to be developed for each variation of the system, considering the yields and characteristics of each process and evaluating the environmental and economic profile using extended and realistic approaches. Currently, some studies approach scenarios for obtaining acids evaluated by LCA, such as getting succinic acid and other value-added products (extract rich in phenolics, oil, calcium tartrate, and tannins) using grape residues as feedstock, where promising results were observed regarding the minimum commercialization value of the acid, which varied between US$ 1.23 and 2.76, being considered by the authors a competitive market price (Ioannidou et al. 2022). When the organic fraction of municipal solid waste was used to evaluate the potential for succinic acid production, the minimum sales value was close to that presented in the study by Ioannidou et al. (2022), ranging from US$1.13 to 2.39, with CO_2 emission reductions of 35% from a system with an input of 1 ton of waste (Ladakis et al. 2022). For the gluconic acid obtained from corn straw, the minimum selling price was U$ 0.404, in a biological process conducted with high solids loading in the hydrolysis step and with sugar yields more significant than 90% (Zhang et al. 2016). The acetic acid production using cottonseed as a feedstock could reduce 156,424 tons of CO_2 for an annual processing capacity of 356,400 tons, which could be reversed in carbon credits that would incorporate into the revenue of a biorefinery (Sayar et al. 2018). These studies are highly relevant to biorefinery projects aimed at organic acid production, thus demonstrating the feasibility of designing biological routes.

Organic acids are relevant products to be considered in 2G biorefineries since they are precursors of several products with high added value. They can be the basis for advances in the sector, with numerous applications and potential to reduce the carbon footprint of production systems, and as an essential product line to sustain a large-scale biorefinery.

3.10. Oligosaccharides

Food as a high-value-added product in biorefineries can drive a significant debate about the possibility of integrating nutrition and chemical chain processes as proposed in the biorefinery engineering project (Solomou et al. 2022; Sheppard et al. 2019). Sheppard et al. (2019) highlight that production costs must be minimized in food and beverage manufacturing since the economic gain margin is usually low. Therefore, integrating processes to obtain different products can be a strategy for cost reduction.

In this sense, the search for alternative biomass sources derived from by-products with potential production of compounds with applications in the food industry has been widely studied. Among these are the functional foods, compounds that contribute to human health by interfering with gastrointestinal microbial activity, especially oligosaccharides, since they can be extracted from structures that make up lignocellulosic biomass (Awasthi et al. 2022). Therefore, numerous types of prebiotics are differentiated by their source and extraction, such as fructooligosaccharides (FOS), glucoligosaccharides (GOS), arabinoxylosaccharides (AXOS), and xylooligosaccharides (XOS), which can be found in various natural sources such as honey, fruits, dairy products, and vegetables, and industrially there is interest in production from lignocellulosic biomass (Pinales-Márquez et al. 2021).

The presence of cellulose and hemicellulose in biomass enables the production of oligosaccharides by fractioning the biomass which allows the production of these compounds in addition to sugars and lignin (Milessi et al. 2021). For example, hemicellulose fractions are usually rich in XOS, manooligosaccharides, and other oligomers since hemicellulose is a heterogeneous polysaccharide. (Xu et al. 2019; Milessi et al. 2021). XOS is among the most exciting oligosaccharides for the economical market, with an estimated growth from USD 64.07 million in 2021 to USD 84.93 million in 2028 (Market Data Forest 2021). Besides having a high value-added and a high capacity for market expansion, it can be a precursor to the production of xylitol, which is a 5-carbon polyol linked to hydroxyl groups that have great value in the food and pharmaceutical industries as an additive and replacement of commonly used sugars (sucrose) (Scapini et al. 2021).

As for any biorefinery development scenario, LCA is essential to analyze the potential for obtaining oligosaccharides. The oligosaccharides' production is usually conducted before the fermentative processes to obtain other compounds, such as biofuels. And the extraction pathway of the matrices that generate these compounds is conducted via pretreatment and hydrolysis. Chemical or biological routes can perform it, the former being more commonly used because of the high yield presented (Bhatia et al. 2019). In these scenarios, it is constantly observed that the recovery of oligosaccharides or other sugars, such as xylose, as the building block of the biorefinery increases the potential for scale-up since the market values of these products are higher compared to other commonly recovered products, such as ethanol (Liu et al. 2021).

In this chapter, the focus is given to oligosaccharides since they can be obtained from structures of lignocellulosic biomass; however, the relevance of designing diverse food industries colocalized to product and energy is recognized in biorefineries. One example is the production of sugar, ethanol, and energy in the Brazilian sugarcane industries. A further expansion of the scenario to other food and chemical products of interest to the economy is expected.

4. Fitting the Biorefineries to the Industry 5.0

The 5th industrial revolution (IR) is a human-centric solution in which robots will be coupled with the human mind and work as collaborators through a brain-machine interface and advances in artificial intelligence. It seems to be a futuristic scenario, but the 5th IR is entering the door. The transition period between industrial revolutions following the exponential advances of the global economy: while the 1st, 2nd, and 3rd lasted centuries, the 4th and 5th IR took only decades to transition (Nahavandi 2019). The transition from industry 3.0 to 4.0 occurred in 2000, and since 2020, industry 5.0 has been coexisting with the promise to have a strong focus and action on the environment and people. It intends to create an ecosystem getting three central protagonists closer: humans, machines, and the environment (Figure 1).

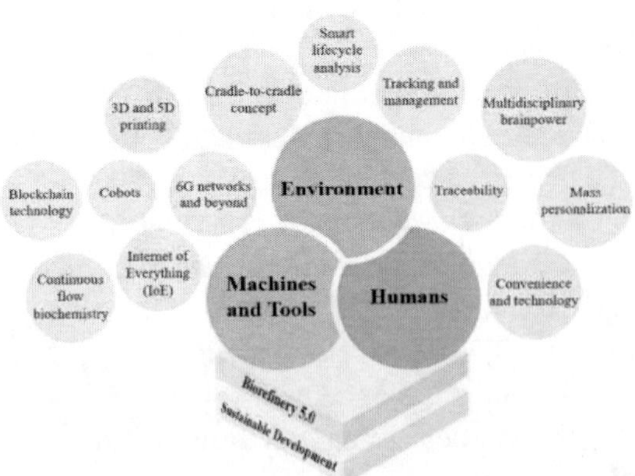

Figure 1. Biorefinery 5.0 and the three central protagonists: humans, machines, and the environment.

Considering all the aspects discussed in the previous topics on biorefineries, it is unquestionable that the future of this sector lies in the 5th IR. To present the importance of industry 5.0 tools and concepts, the following topics address the three main protagonists and how each of them can contribute to creating biorefinery 5.0.

4.1. Humans

Although the three protagonists are equally important, the 5th IR has been defined as a human-centric solution; therefore, in this scenario of economic development, it must be the products and processes of biorefineries that will be focused on people's demands. The biobased products are already well grounded on consumer trends such as health, sustainability, waste reduction, and new flavors. But these trends can be considered the basic ones since society 5.0 demands advanced consumption experiences (Onday 2019). The behavior of deciding on a specific brand and its products has undergone a significant change. The industry needs to readjust its concepts to deliver personalized products to consumers. Personalization is the great highlight of industry 5.0 when considering what the consumer expects from the future. While industry 4.0 is focused on mass production, industry 5.0 is focused on mass personalization, that is, in delivering personalized products according to the customer's taste and needs. Another aspect that has gained market attention (when considering the entire chain surrounding a commodity) is traceability, i.e., the biobased products must have an easily accessible database containing information on where each part of the final product comes from and how it is produced (Maddikunta et al. 2022; Nahavandi 2019).

The consumer analysis and choice start to permeate beyond the possession desires: society 5.0 expects to have a personal identification with what it consumes. There is also the emerging aspect of belonging to a community, which has advanced with the virtual social networks and has made people increasingly influenced by what other consumes. In this way, the consumer aspect of a person becomes closely related to personal values (Onday 2019). In other words, "you are what you consume" becomes a philosophy of life, and personalization and traceability become fundamental to helping consumers choose their lifestyles. It should be noted that the social media rise has caused an acceleration in product and service obsolescence and in the speed at which consumers expect to receive information. Therefore, biorefinery 5.0 success lies in its constant reinvention and the creation of simple interfaces to easily access information.

Also, in the human aspect of biorefinery 5.0, the importance of qualified and educated employees should be highlighted. For this interconnection between consumer and industry to work correctly, the human brain must interpret all machine-generated data (Polat and Erkollar 2021). The 5th IR brings back human workers to the factory floors and understands that human brainpower and creativity are valuable to increase process efficiency by combining workflows with machines (Nahavandi 2019). In biorefinery 5.0, humans and robots will be equally important and work synergistically to meet consumer demands and expand the biobased market. The required advances to insert biorefineries in this increasingly competitive market will only be possible using the machines and tools presented in the following topic.

4.2. Machines and Tools

Biorefineries have the power to meet society's 5.0 demands because their processes and products are strongly aligned with the social and environmental principles that people expect from industries. With the increasing speed, quality, and quantity of these demands, we can summarize what is needed for the future of biorefineries in one word: the internet. All machines and tools of industry 5.0 must be somehow connected to the virtual universe, which has the potential to connect industry with supercomputers and interconnect multiple industries in the biorefinery sector with each other. The biorefinery network will produce biofuels, nanomaterials, organic acids, functional foods, bioplastics, and all other possible products from biobased materials. To fully exploit the power of the internet and the virtual world, the 5G technology has been essential in the current industry 4.0. Still, industry 5.0 will need to go further and use 6G networks and beyond. Through the 6G, it will be possible to deliver low latency, smart resource management, ultra-high data rates, high-quality services, high energy efficiency, high traffic capacity, and extensive internet of everything (IoE) infrastructure (Maddikunta et al. 2022). If all these factors are used in favor of the development of biorefineries, the potential of this sector can be more quickly established in the economic market.

As already mentioned, the data is valuable as raw material for biorefinery 5.0, although it is unuseful without analytics processes. The large and diverse data collected from field, factory, market, and consumers can benefit the biorefinery network after being analyzed by 5th IR tools such as machine learning, artificial intelligence, data mining, data fusion, edge computing, etc. The importance of organized and effective extensive data analysis will help the biorefinery make better decisions and real-time monitoring of their processes, production, sales, and so on (Yarali 2021; Maddikunta et al. 2022).

The equipment and processes used by biorefineries can be optimized, operated, and automatized because of the big data and correlated industry 5.0 tools (e.g., mathematical modeling, experimental design, statistical treatment, etc.). During the construction of bioreactors and associated equipment (e.g., screws, gears, connectors, pipes, sensors, etc.), another industry 5.0 concept can be used: smart additive manufacturing by 3D and 5D printing. This approach to designing and producing bioreactors and equipment is a layer-by-layer manufacturing method that can accelerate mathematical modeling and phenomena prediction, improve productivity and give sustainability to biorefineries (Neyt and Riley 2021; Maddikunta et al. 2022). It also helps the process scalability and cost reduction, especially important for the pharmaceutical and food industries operating their bioprocesses with single-use reactors.

These changes in reactor engineering concepts will benefit biorefinery 5.0 and open an opportunity to adopt continuous flow biochemistry, integrating advanced computer and IoE technologies for multistep synthesis under flow conditions. Some examples of continuous flow chemistry practices are algorithmic chemistry, flow automation allowing several reactions on the same system, production of new and known target compounds efficiently, digital cameras connected to computer vision algorithms, and purification/ downstream processing technologies in flow, among others (Neyt and Riley 2021). The application of such tools promises to merge biochemistry, engineering, and informatics.

4.3. Environment

Biorefineries significantly contribute to the industry's sustainable development, as this chapter has extensively discussed. In addition, it has great potential to meet social demands through the machines and tools of the 5th IR presented in the previous topic. There is nothing more to discuss the environment, but the concepts of industry 5.0 indicate that there is. For Xu, Lu, and Li (2021) and Nahavandi (2019), the current industry 4.0 failed to balance the relationship between man, machine, and the environment. While the 4th IR achieved efficient costs reduction of production (10–30%), logistic

(10–30%), and quality management (10–20%), the 5th IR is required to reduce human and environmental costs. That is a human-centric revolution that will change the focus from things to personal values, taking sustainability and resilience as interconnected core values. In this regard, biorefineries and the environment have as great allies the tools and concepts of the 5th IR.

When it comes to environmental management and smart lifecycle analysis, big data becomes a feasible basis for monitoring an entire production chain and managing what happens throughout all life stages, assisting in decision-making to remediate eventual problems, and providing information to address contaminant emissions in the air, soil, and water, or even reduce the emissions. Taking plant-based food as an example, obtaining and storing data can help predict how the soil quality is where the plant was grown and when the soil will need intervention. It will be possible to monitor the quantity and quality of water used in the field and the need to intervene in the water supply source. In the production process at the factory, water monitoring will also define how much water is being consumed, what is the yield of the process in terms of water use and how are the quality and quantity of generated wastewater. Based on these data, the quantity and quality of air emissions can be controlled, as well as solid waste. These data can feed the treatment processes of the contaminants in quantitative and qualitative terms, improving predictability and real-time action. After leaving the factory, the continued data collection will assist in the environmental management of the post-consumer product, given that possible environmental damage may still exist. It is the biorefinery's responsibility to deal with all associated impacts.

The environmental tracking of bioproducts and bioprocesses allows biorefineries to apply the cradle-to-cradle concept, which aligns with the SDGs, the circular economy, and industry 5.0. In the cradle-to-cradle lifecycle assessment, it is understood that a product and all its components used and generated during and after production can return to industrial activity as feedstock. Once again, big data, artificial intelligence, mathematical modeling, robots, and all other tools associated with industry 5.0 are great allies of biorefineries for applying emerging concepts related to the environment. The reduction/elimination of exhaustible materials consumption is an essential objective of biorefineries, and the industry 5.0 tools can make it possible.

For biorefinery 5.0 to become a reality, some central points must be the focus of industries when designing a biobased product: (i) what are the consumers' and stakeholders' values linked to the bioproduct, (ii) what are the human and environmental costs linked to the bioproduct, and (iii) how this bioproduct fits into the ecosystem of other biorefineries. The answers to these points must be clear for all materials and processes involved in the production, from the field to the post-sale. The industry 5.0 tools and concepts presented in this chapter are critical to meeting these points.

5. What If We Disregard the Potential of Biorefineries?

Biorefineries can be an essential strategy for the transition to a circular economy, linked mainly to the benefit of using plant and waste biomasses as sources of products and energy. In this way, the extraction of natural resources is reduced, decreasing the dependency on fossil resources and advancing food security, having in mind the goal of reducing environmental impacts by reducing waste and effluents, reducing gas emissions, and drinking water consumption (Igbokwe et al. 2022; European Environment Agency 2018).

Alzaabi and Mezher (2021) provide perspectives on the increased demand for the axis that encompasses water, energy, and food by 2050 and state that the numbers must be doubled. For 2030, the increases are 40%, 50%, and 35%, respectively. The population increase is 80 million people yearly, with a 1% increase in water consumption (Keyhanpour, Musavi Jahromi, and Ebrahimi 2021). Consequently, agricultural production will be expanded by up to 60 percent by 2030, consuming more energy and land. This increased production demand is also associated with industrial processes and animal feed (FAO 2011; 2017b). In this increase in consumption, the SDGs designed for 2030 are indispensable to link issues between food, society, and natural resource management, since they will not succeed if treated individually (FAO 2017a).

In predicting population and production growth, there must be an expansion in insight and understanding beyond, considering that more waste will be generated, more water and arable will be needed, and if efforts are not directed to alternative resources, exclusive dependence on fossil resources could be the big problem of the century. As mentioned above, the numbers regarding the extraction and consumption of natural resources are growing, and new processing techniques must be improved to solve this challenge. In this sense, biomass from different processes and productions can guarantee innovation and energy production through biorefineries, which aim to integrate strategies to obtain food, chemical products, and energy (Velvizhi et al. 2022).

But a question must be answered: is it possible that the biorefinery reaches 0 waste? As seen above, the use of waste in biorefineries to produce new products and biofuels is gaining increasing attention since they are complex systems designed to reduce environmental impacts, reduce dependence on fossil fuels, and generate capital returns.

Ideally, when designing biorefineries, the focus is on the integral use of the feedstock. However, in practice, one deals with a complex system, since besides being thought in the environmental scope, it must follow the economic and social development. To extract the most significant quantity of products, in an optimized system, with lower water consumption, less generation of effluents and residues, and potential for expansion in the economic market.

The biorefinery concept is discussed to improve economic development for lower impacts generated by oil-based products and reduce the amount of waste in landfills (Liu et al. 2021). Research indicates that in 1G and 2G biorefineries, greenhouse gas emissions are reduced by up to 80% compared to standard fuels (Pinto et al. 2022). But one of the biggest problems linked to biorefineries is the high water consumption. Medina et al. (2021) indicate a water consumption of 38.7 $ton.h^{-1}$. For this to be reversed, one of the proposals is a water recovery unit allied to the other units, causing the water to circulate through the processes, decreasing the excessive intake.

There is mutual agreement that the environmental impacts of a biorefinery's implementation are so large and significant that when the associated challenges (i.e., the complexity of system integration, seasonal input of biomass) are put into perspective, one realizes that they are malleable and can be mitigated.

In addition to expanding markets, biorefineries create opportunities for countries to achieve economic development sustainably. Developing countries have many benefits in implementing biorefineries, bringing possibilities to expand markets, supply in short chains, promoting economic growth and the circular economy. Along with this, there is the generation of new jobs, social equity, and the expansion of environmental education (Igbokwe et al. 2022). For example, Indonesia and Malaysia produce tons of waste after palm oil extraction that would be converted into high-value products (200 biochemicals and biopolymers). A study by Lestari, Zvinavashe, and Sanders (2015) highlights the use of Jatropha seeds, grown in countries like Zimbabwe, Tanzania, Mali, Indonesia, and Holland, where the cake generated as a byproduct from the manufacture of biodiesel can develop rural and industrial products. The plantation of 1 hectare of this typical plant can create income and jobs for these countries, focusing on exporting products to generate revenue for the country and attractiveness to the rural community.

By 2030, Europe has the goal that up to 30% of products considered chemicals come from processes based on a circular economy. The biggest bottlenecks are in obtaining biomass in large production quantities and the price of the components used during production (Zeng 2019). The progress or failure of implementation will also depend on public policy, and more significant political incentives will lead to new research and proposals.

Finally, boosting the development of biorefineries also goes a long way towards achieving the SDGs. When it comes to people below the poverty line and populations living without access to essential services, biorefineries generate employment opportunities and contribute by making electricity available to rural and urban areas. Also, it can reduce food loss using waste customarily disposed of in inappropriate places as a source for producing value-added products (Solarte-Toro and Alzate 2021).

Conclusion

Biorefineries as a strategy to achieve a more sustainable economy in the long term is a promising way forward. With the expansions in the biotechnological market and the products obtained by biomass, the development of integrated systems tends to be more and more explored. Bio-based products for different areas (food, chemical, pharmaceutical, etc.) tend to expand the allocation in the economic market and achieve more competitive prices than non-renewable products. And with the advancement of technologies and the introduction of Industry 5.0, biorefineries can support economic and social growth, moving towards the SDGs and helping to reduce environmental problems.

References

Aguiar, Jéssica de, Thalita Jessika Bondancia, Pedro Ivo C. Claro, Luiz Henrique C. Mattoso, Cristiane S. Farinas, and José Manoel Marconcini. 2020. "Enzymatic Deconstruction of Sugarcane Bagasse and Straw to Obtain Cellulose Nanomaterials." *ACS Sustainable Chemistry & Engineering* 8 (5): 2287–99. https://doi.org/10.1021/acssuschemeng.9b06806.

Alzaabi, Mohammed S.M. Alasam, and Toufic Mezher. 2021. "Analyzing Existing UAE National Water, Energy and Food Nexus Related Strategies." *Renewable and Sustainable Energy Reviews* 144 (July): 111031. https://doi.org/10.1016/j.rser.2021.111031.

Amaro Bittencourt, Gustavo, Luciana Porto de Souza Vandenberghe, Kim Valladares-Diestra, Leonardo Wedderhoff Herrmann, Ariane Fátima Murawski de Mello, Zulma Sarmiento Vásquez, Susan Grace Karp, and Carlos Ricardo Soccol. 2021. "Soybean Hulls as Carbohydrate Feedstock for Medium to High-Value Biomolecule Production in Biorefineries: A Review." *Bioresource Technology* 339 (May). https://doi.org/10.1016/j.biortech.2021.125594.

Arutselvan, Chithirai, Felix Lewis, G Ramya, Nguyen Thuy, Lan Chi, Arivalagan Pugazhendhi, and Nooruddin Thajuddin. 2022. "Review on Wastewater Treatment by Microalgae in Different Cultivation Systems and Its Importance in Biodiesel Production." *Fuel* 324 (April). https://doi.org/https://doi.org/10.1016/j.fuel.2022.124623.

Astolfi, Angela Luiza, Alan Rempel, Vítor Augusto Farina Cavanhi, Maycon Alves, Kricelle Mosquera Deamici, Luciane Maria Colla, and Jorge Alberto Viera Costa. 2020. "Simultaneous Saccharification and Fermentation of Spirulina Sp. and Corn Starch for the Production of Bioethanol and Obtaining Biopeptides with High Antioxidant Activity." *Bioresource Technology* 301 (December 2019). https://doi.org/10.1016/j.biortech.2019.122698.

Awasthi, Mukesh Kumar, Ayon Tarafdar, Vivek Kumar Gaur, K. Amulya, Vivek Narisetty, Dheeraj Kumar Yadav, Raveendran Sindhu, et al. 2022. "Emerging Trends of Microbial Technology for the Production of Oligosaccharides from Biowaste and Their Potential Application as Prebiotic." *International Journal of Food Microbiology* 368 (May): 109610. https://doi.org/10.1016/j.ijfoodmicro.2022.109610.

Baig, Nadeem, Irshad Kammakakam, and Wail Falath. 2021. "Nanomaterials: A Review of Synthesis Methods, Properties, Recent Progress, and Challenges." *Materials Advances* 2 (6): 1821–71. https://doi.org/10.1039/D0MA00807A.

Banu, J. Rajesh, G. Ginni, S. Kavitha, R. Yukesh Kannah, S. Adish Kumar, Shashi Kant Bhatia, and Gopalakrishnan Kumar. 2021. "Integrated Biorefinery Routes of Biohydrogen: Possible Utilization of Acidogenic Fermentative Effluent." *Bioresource Technology* 319 (January): 124241. https://doi.org/10.1016/J.BIORTECH.2020.124241.

Banu, J. Rajesh, S. Kavitha, R. Yukesh Kannah, Rahul R. Bhosale, and Gopalakrishnan Kumar. 2020. "Industrial Wastewater to Biohydrogen: Possibilities towards Successful Biorefinery Route." *Bioresource Technology* 298 (February): 122378. https://doi.org/10.1016/J.BIORTECH.2019.122378.

Barreiros, Thaís, André Young, Raquel Cavalcante, and Eduardo Queiroz. 2020. "Impact of Biodiesel Production on a Soybean Biorefinery." *Renewable Energy* 159: 1066–83. https://doi.org/10.1016/j.renene.2020.06.064.

Battista, Federico, Nicola Frison, Paolo Pavan, Cristina Cavinato, Marco Gottardo, Francesco Fatone, Anna L. Eusebi, et al. 2020. "Food Wastes and Sewage Sludge as Feedstock for an Urban Biorefinery Producing Biofuels and Added-Value Bioproducts." *Journal of Chemical Technology and Biotechnology* 95 (2): 328–38. https://doi.org/10.1002/jctb.6096.

BBC Research. 2021. "*Global Markets and Technologies for Bioplastics*." https://www.bccresearch.com/market-research/plastics/global-markets-and-technologies-for-bioplastics.html#:~:text=Report Highlights,period of 2021 to 2026.

Bhatia, Latika, Ashutosh Sharma, Rakesh K. Bachheti, and Anuj K. Chandel. 2019. "Lignocellulose Derived Functional Oligosaccharides: Production, Properties, and Health Benefits." *Preparative Biochemistry and Biotechnology* 49 (8): 744–58. https://doi.org/10.1080/10826068.2019.1608446.

Bitonto, Luigi di, Stefano Todisco, Vito Gallo, and Carlo Pastore. 2020. "Urban Sewage Scum and Primary Sludge as Profitable Sources of Biodiesel and Biolubricants of New Generation." *Bioresource Technology Reports* 9 (January). https://doi.org/10.1016/j.biteb.2020.100382.

Bohn, Letícia Renata, Aline Perin Dresch, Matheus Cavali, Ana Carolina Giacomelli Vargas, Jaíne Flach Führ, Siumar Pedro Tironi, Odinei Fogolari, Guilherme Martinez Mibielli, Sérgio Luiz Alves Junior, and João Paulo Bender. 2021. "Alkaline Pretreatment and Enzymatic Hydrolysis of Corn Stover for Bioethanol Production." *Research, Society and Development* 2021 (2525–3409): 1–13. https://doi.org/https://doi.org/10.33448/rsd-v10i11.18914.

Bošnjaković, Mladen, and Nazaruddin Sinaga. 2020. "The Perspective of Large-Scale Production of Algae Biodiesel." *Applied Sciences (Switzerland)* 10 (22): 1–26. https://doi.org/10.3390/app10228181.

Bruins, Marieke E., and Johan P.M. Sanders. 2012. "Small-Scale Processing of Biomass for Biorefinery." *Biofuels, Bioproducts and Biorefining* 6 (2): 135–45. https://doi.org/10.1002/bbb.1319.

Bukhari, Nurul Adela, Jamaliah Md Jahim, Soh Kheang Loh, Abu Bakar Nasrin, Shuhaida Harun, and Peer Mohamed Abdul. 2020. "Organic Acid Pretreatment of Oil Palm Trunk Biomass for Succinic Acid Production." *Waste and Biomass Valorization* 11 (10): 5549–59. https://doi.org/10.1007/s12649-020-00953-2.

Cai, Di, Jieyi Wen, Yan Zhuang, Tianhao Huang, Zhihao Si, Peiyong Qin, and Huidong Chen. 2022. "Review of Alternative Technologies for Acetone-Butanol-Ethanol Separation: Principles, State-of-the-Art, and Development Trends." *Separation and Purification Technology*, May, 121244. https://doi.org/10.1016/J.SEPPUR.2022.121244.

Calicioglu, Özgül, and Anne Bogdanski. 2021. "Linking the Bioeconomy to the 2030 Sustainable Development Agenda: Can SDG Indicators Be Used to Monitor Progress towards a Sustainable Bioeconomy?" *New Biotechnology* 61 (March): 40–49. https://doi.org/10.1016/j.nbt.2020.10.010.

Castro-Aguirre, E., F. Iñiguez-Franco, H. Samsudin, X. Fang, and R. Auras. 2016. "Poly(Lactic Acid)—Mass Production, Processing, Industrial Applications, and End of Life." *Advanced Drug Delivery Reviews* 107 (December): 333–66. https://doi.org/10.1016/j.addr.2016.03.010.

Cavali, Matheus, Auriane Bueno, Ana P. Fagundes, Wagner L. Priamo, Denise Bilibio, Guilherme M. Mibielli, João H.C. Wancura, João P. Bender, and J. Vladimir Oliveira. 2020. "Liquid Lipase-Mediated Production of Biodiesel from Agroindustrial Waste." *Biocatalysis and Agricultural Biotechnology* 30 (November). https://doi.org/10.1016/j.bcab.2020.101864.

Cavali, Matheus, Nelson Libardi Junior, Rodrigo De Almeida Mohedano, Paulo Belli Filho, Rejane Helena, Armando Borges, and De Castilhos Junior. 2022. "Biochar and Hydrochar in the Context of Anaerobic Digestion for a Circular Approach: An

Overview." *Science of the Total Environment* 822: 153614. https://doi.org/http://dx.doi.org/10.1016/j.scitotenv.2022.153614.

Cavali, Matheus, Carlos Ricardo Soccol, Débora Tavares, Luis Alberto Zevallos Torres, Valcineide Oliveira de Andrade Tanobe, Arion Zandoná Filho, and Adenise Lorenci Woiciechowski. 2021. "Valorization of Lignin from Pine (Pinus Spp.) Residual Sawdust: Antioxidant Activity and Application in the Green Synthesis of Silver Nanoparticles for Antibacterial Purpose." *Biomass Conversion and Biorefinery*, no. 0123456789. https://doi.org/10.1007/s13399-021-01940-w.

Chen, Guo-Qiang. 2009. "A Microbial Polyhydroxyalkanoates (PHA) Based Bio- and Materials Industry." *Chemical Society Reviews* 38 (8): 2434. https://doi.org/10.1039/b812677c.

Cheng, Dongle, Huu Hao Ngo, Wenshan Guo, Soon Woong Chang, Dinh Duc Nguyen, Xuan Thanh Bui, Wei Wei, Bingjie Ni, Sunita Varjani, and Ngoc Bich Hoang. 2022. "Enhanced Photo-Fermentative Biohydrogen Production from Biowastes: An Overview." *Bioresource Technology* 357 (August): 127341. https://doi.org/10.1016/J.BIORTECH.2022.127341.

Cherubini, Francesco. 2010. "The Biorefinery Concept: Using Biomass Instead of Oil for Producing Energy and Chemicals." *Energy Conversion and Management* 51 (7): 1412–21. https://doi.org/10.1016/j.enconman.2010.01.015.

Chia, Shir Reen, Kit Wayne Chew, Pau Loke Show, Yee Jiun Yap, Hwai Chyuan Ong, Tau Chuan Ling, and Jo Shu Chang. 2018. "Analysis of Economic and Environmental Aspects of Microalgae Biorefinery for Biofuels Production: A Review." *Biotechnology Journal* 13 (6): 1–10. https://doi.org/10.1002/biot.201700618.

Choe, Bomin, Shinje Lee, Hyunjun Lee, Jinwon Lee, Hankwon Lim, and Wangyun Won. 2021. "Integrated Strategy for Coproducing Bioethanol and Adipic Acid from Lignocellulosic Biomass." *Journal of Cleaner Production* 311 (June). https://doi.org/10.1016/j.jclepro.2021.127849.

Ciriminna, Rosaria, Francesco Meneguzzo, Riccardo Delisi, and Mario Pagliaro. 2017. "Citric Acid: Emerging Applications of Key Biotechnology Industrial Product." *Chemistry Central Journal* 11 (1): 22. https://doi.org/10.1186/s13065-017-0251-y.

Conteratto, Caroline, Felipe Dalzotto Artuzo, Omar Inácio Benedetti Santos, and Edson Talamini. 2021. "Biorefinery: A Comprehensive Concept for the Sociotechnical Transition toward Bioeconomy." *Renewable and Sustainable Energy Reviews* 151 (November): 111527. https://doi.org/10.1016/j.rser.2021.111527.

Crutchik, Dafne, Oscar Franchi, Luis Caminos, David Jeison, Marisol Belmonte, Alba Pedrouso, Angeles Val del Rio, Anuska Mosquera-Corral, and José Luis Campos. 2020. "Polyhydroxyalkanoates (PHAs) Production: A Feasible Economic Option for the Treatment of Sewage Sludge in Municipal Wastewater Treatment Plants?" *Water* 12 (4): 1118. https://doi.org/10.3390/w12041118.

D'Amato, D., and J. Korhonen. 2021. "Integrating the Green Economy, Circular Economy and Bioeconomy in a Strategic Sustainability Framework." *Ecological Economics* 188 (October): 107143. https://doi.org/10.1016/j.ecolecon.2021.107143.

Daful, Asfaw Gezae, and Johann F. Görgens. 2017. "Techno-Economic Analysis and Environmental Impact Assessment of Lignocellulosic Lactic Acid Production."

Chemical Engineering Science 162 (April): 53–65. https://doi.org/10.1016/j.ces.2016.12.054.

Deuber, Raquel de Souza, Daniel Santos Fernandes, Jéssica Marcon Bressanin, Jamison Watson, Mateus Ferreira Chagas, Antonio Bonomi, Leonardo Vasconcelos Fregolente, and Marcos Djun Barbosa Watanabe. 2021. "Techno-Economic Assessment of HTL Integration to the Brazilian Sugarcane Industry: An Evaluation of Different Scenarios." *Industrial Crops and Products* 173 (March). https://doi.org/10.1016/j.indcrop.2021.114139.

Dezam, A.P.G., V.M. Vasconcellos, P.T. Lacava, and C.S. Farinas. 2017. "Microbial Production of Organic Acids by Endophytic Fungi." *Biocatalysis and Agricultural Biotechnology* 11 (July): 282–87. https://doi.org/10.1016/j.bcab.2017.08.001.

Duan, Yumin, Ayon Tarafdar, Vinay Kumar, Prabakaran Ganeshan, Karthik Rajendran, Balendu Shekhar Giri, Ricardo Gómez-García, et al. 2022. "Sustainable Biorefinery Approaches towards Circular Economy for Conversion of Biowaste to Value Added Materials and Future Perspectives." *Fuel* 325 (October): 124846. https://doi.org/10.1016/j.fuel.2022.124846.

Dzulkarnain, Eka Latiffah Nadia, Jemilatu Omuwa Audu, Wan Rosmiza Zana Wan Dagang, and Mohd Firdaus Abdul-Wahab. 2022. "Microbiomes of Biohydrogen Production from Dark Fermentation of Industrial Wastes: Current Trends, Advanced Tools and Future Outlook." *Bioresources and Bioprocessing* 9 (1). https://doi.org/10.1186/s40643-022-00504-8.

El-Qelish, Mohamed, Gamal K. Hassan, Sebastian Leaper, Paolo Dessì, and Ahmed Abdel-Karim. 2022. "Membrane-Based Technologies for Biohydrogen Production: A Review." *Journal of Environmental Management* 316 (August): 115239. https://doi.org/10.1016/J.JENVMAN.2022.115239.

Eng Sánchez, Felipe, Lucas Tadeu Fuess, Guilherme Soares Cavalcante, Maria Ângela Talarico Adorno, and Marcelo Zaiat. 2021. "Value-Added Soluble Metabolite Production from Sugarcane Vinasse within the Carboxylate Platform: An Application of the Anaerobic Biorefinery beyond Biogas Production." *Fuel* 286 (February): 119378. https://doi.org/10.1016/J.FUEL.2020.119378.

European Bioplastic. 2021. "*Bioplastics Market Data 2021*." https://docs.european-bioplastics.org/publications/market_data/Report_Bioplastics_Market_Data_2021_short_version.pdf.

European Environment Agency. 2018. *The Circular Economy and the Bioeconomy: Partners in Sustainability*. Luxembourg: Publications Office of the European Union. https://doi.org/10.2800/02937.

Fact.MR. 2021. "*Organic Acids Market - Global Insights 2021-2031*." https://www.factmr.com/report/4285/organic-acids-market.

FAO. 2011. "*Energy-Smart Food for People and Climate*." https://www.fao.org/3/i2454e/i2454e.pdf.

———. 2017a. "F*ood and Agriculture: Driving Action across the 2030 Agenda for Sustainable Development*." https://www.fao.org/3/i7454e/i7454e.pdf.

———. 2017b. "*The Future of Food and Agriculture- Trends and Challenges*." https://www.fao.org/3/i6881e/i6881e.pdf.

Ferreira-Filipe, Diogo A., Ana Paço, Armando C. Duarte, Teresa Rocha-Santos, and Ana L. Patrício Silva. 2021. "Are Biobased Plastics Green Alternatives?—A Critical Review." *International Journal of Environmental Research and Public Health* 18 (15): 7729. https://doi.org/10.3390/ijerph18157729.

Food and Agriculture Organization, FAO. 2020. *OECD-FAO Agricultural Outlook 2020-2029*. OECD Publishing Paris, /Food and Agriculture Organization of the United Nations, Rome.

Formann, Steffi, Alena Hahn, Leandro Janke, Walter Stinner, Heike Sträuber, Washington Logroño, and Marcell Nikolausz. 2020. "Beyond Sugar and Ethanol Production: Value Generation Opportunities Through Sugarcane Residues." *Frontiers in Energy Research*. https://doi.org/10.3389/fenrg.2020.579577.

GBS. 2020. "*Global Bioeconomy Summit Conference 2020*." https://gbs2020.net/conference-report/.

Gerrior, Daniel, Khashayar Delsoz Bahri, Azadeh Kermanshahi-pour, Mathew J. Eckelman, and Satinder Kaur Brar. 2022. "Life Cycle Assessment and Techno-Economic Analysis of a Novel Closed Loop Corn Ethanol Biorefinery." *Sustainable Production and Consumption* 30: 359–76. https://doi.org/10.1016/j.spc.2021.12.007.

Ghomi, Erfan Rezvani, Fatemeh Khosravi, Ali Saedi Saedi Ardahaei, Yunqian Dai, Rasoul Esmaeely Neisiany, Firoozeh Foroughi, Min Wu, Oisik Das, and Seeram Ramakrishna. 2021. "The Life Cycle Assessment for Polylactic Acid (PLA) to Make It a Low-Carbon Material." *Polymers* 13 (11): 1854. https://doi.org/10.3390/polym13111854.

GMI. 2022. "*Nanomaterials Market Size, Industry Analysis Report, Regional Outlook, Application Development Potential, Price Trend, Competitive Market Share & Forecast, 2022 – 2028*." https://www.gminsights.com/industry-analysis/nanomaterials-market.

Guilayn, F., M. Rouez, M. Crest, D. Patureau, and J. Jimenez. 2020. Valorization of Digestates from Urban or Centralized Biogas Plants: A Critical Review. *Reviews in Environmental Science and Biotechnology*. Vol. 19. https://doi.org/10.1007/s11157-020-09531-3.

Gwalwanshi, Manoj, Rajesh Kumar, and Manish Kumar Chauhan. 2022. "A Review on Butanol Properties, Production and Its Application in Internal Combustion Engines." *Materials Today: Proceedings*, April. https://doi.org/10.1016/J.MATPR.2022.04.573.

Hassan, Shady S., Gwilym A. Williams, and Amit K. Jaiswal. 2019. "Moving towards the Second Generation of Lignocellulosic Biorefineries in the EU: Drivers, Challenges, and Opportunities." *Renewable and Sustainable Energy Reviews* 101 (March): 590–99. https://doi.org/10.1016/j.rser.2018.11.041.

Hou, Li, Ling Liu, Hongfei Zhang, Lin Zhang, Lan Zhang, Jian Zhang, Qiang Gao, and Depei Wang. 2018. "Functional Analysis of the Mitochondrial Alternative Oxidase Gene (Aox1) from Aspergillus Niger CGMCC 10142 and Its Effects on Citric Acid Production." *Applied Microbiology and Biotechnology* 102 (18): 7981–95. https://doi.org/10.1007/s00253-018-9197-9.

Igbokwe, Victor C., Flora N. Ezugworie, Chukwudi O. Onwosi, Godwin O. Aliyu, and Chinonye J. Obi. 2022. "Biochemical Biorefinery: A Low-Cost and Non-Waste

Concept for Promoting Sustainable Circular Bioeconomy." *Journal of Environmental Management* 305 (March): 114333. https://doi.org/10.1016/j.jenvman.2021.114333.

International Energy Agency. 2020. "Outlook for Biogas and Biomethane - Prospects for Organic Growth." *International Energy Agency (IEA) - Bioenergy*. https://doi.org/10.1787/040c8cd2-en.

———. 2022. "*Hydrogen - Analysis*."

Ioannidou, Sofia Maria, Katiana Filippi, Ioannis K. Kookos, Apostolis Koutinas, and Dimitrios Ladakis. 2022. "Techno-Economic Evaluation and Life Cycle Assessment of a Biorefinery Using Winery Waste Streams for the Production of Succinic Acid and Value-Added Co-Products." *Bioresource Technology* 348 (March): 126295. https://doi.org/10.1016/j.biortech.2021.126295.

Iqbal, Khushboo, Abhishek Saxena, Priyanshi Pande, Archana Tiwari, Naveen Chandra Joshi, Ajit Varma, and Arti Mishra. 2022. "Microalgae-Bacterial Granular Consortium: Striding towards Sustainable Production of Biohydrogen Coupled with Wastewater Treatment." *Bioresource Technology* 354 (June): 127203. https://doi.org/10.1016/J.BIORTECH.2022.127203.

Jain, Sarika, David Newman, Ange Nzihou, Harmen Dekker, Pharoah Le Feuvre, Hannah Richter, Frederic Gobe, Charlotte Morton, and Rebecca Thompson. 2019. "Global Potential of Biogas." *World Biogas Association*.

Joglekar, Saurabh N., Pranav D. Pathak, Sachin A. Mandavgane, and Bhaskar D. Kulkarni. 2019. "Process of Fruit Peel Waste Biorefinery: A Case Study of Citrus Waste Biorefinery, Its Environmental Impacts and Recommendations." *Environmental Science and Pollution Research* 26 (34): 34713–22. https://doi.org/10.1007/s11356-019-04196-0.

Kahar, Prihardi, Nova Rachmadona, Radityo Pangestu, Rendi Palar, Deddy Triyono Nugroho Adi, Ario Betha Juanssilfero, Yopi, Immanuel Manurung, Shinji Hama, and Chiaki Ogino. 2022. "An Integrated Biorefinery Strategy for the Utilization of Palm-Oil Wastes." *Bioresource Technology* 344 (January): 126266. https://doi.org/10.1016/j.biortech.2021.126266.

Katakojwala, Ranaprathap, and S. Venkata Mohan. 2022. "Multi-Product Biorefinery with Sugarcane Bagasse: Process Development for Nanocellulose, Lignin and Biohydrogen Production and Lifecycle Analysis." *Chemical Engineering Journal* 446 (October): 137233. https://doi.org/10.1016/J.CEJ.2022.137233.

Keyhanpour, Mohammad Javad, Seyed Habib Musavi Jahromi, and Hossein Ebrahimi. 2021. "System Dynamics Model of Sustainable Water Resources Management Using the Nexus Water-Food-Energy Approach." *Ain Shams Engineering Journal* 12 (2): 1267–81. https://doi.org/10.1016/j.asej.2020.07.029.

Khan, Bahram, Muhammad Bilal Khan Niazi, Ghufrana Samin, and Zaib Jahan. 2017. "Thermoplastic Starch: A Possible Biodegradable Food Packaging Material-A Review." *Journal of Food Process Engineering* 40 (3): e12447. https://doi.org/10.1111/jfpe.12447.

Khatami, Kasra, Mariel Perez-Zabaleta, Isaac Owusu-Agyeman, and Zeynep Cetecioglu. 2021. "Waste to Bioplastics: How Close Are We to Sustainable Polyhydroxyalkanoates Production?" *Waste Management* 119 (January): 374–88. https://doi.org/10.1016/j.wasman.2020.10.008.

Koch, Daniel, Marco Paul, Stefan Beisl, Anton Friedl, and Bettina Mihalyi. 2020. "Life Cycle Assessment of a Lignin Nanoparticle Biorefinery: Decision Support for Its Process Development." *Journal of Cleaner Production* 245 (February): 118760. https://doi.org/10.1016/j.jclepro.2019.118760.

Kuenz, Anja, and Susan Krull. 2018. "Biotechnological Production of Itaconic Acid—Things You Have to Know." *Applied Microbiology and Biotechnology* 102 (9): 3901–14. https://doi.org/10.1007/s00253-018-8895-7.

Ladakis, Dimitrios, Eleni Stylianou, Sofia-Maria Ioannidou, Apostolis Koutinas, and Chrysanthi Pateraki. 2022. "Biorefinery Development, Techno-Economic Evaluation and Environmental Impact Analysis for the Conversion of the Organic Fraction of Municipal Solid Waste into Succinic Acid and Value-Added Fractions." *Bioresource Technology* 354 (June): 127172. https://doi.org/10.1016/j.biortech.2022.127172.

Larnaudie, Valeria, Mario Daniel Ferrari, and Claudia Lareo. 2021. "Life Cycle Assessment of Ethanol Produced in a Biorefinery from Liquid Hot Water Pretreated Switchgrass." *Renewable Energy* 176: 606–16. https://doi.org/10.1016/j.renene.2021.05.094.

Le, Tan M., Uyen P.N. Tran, Yen H.P. Duong, Quan D. Nguyen, Viet T. Tran, Phong T. Mai, and Phung K. Le. 2022. "Sustainable Bioethanol and Value-Added Chemicals Production from Paddy Residues at Pilot Scale." *Clean Technologies and Environmental Policy* 24 (1): 185–97. https://doi.org/10.1007/s10098-021-02097-w.

Lena, Gabriella Di, Jose Sanchez Del Pulgar, Massimo Lucarini, Alessandra Durazzo, Petra Ondrejíčková, Florin Oancea, Rodica Mihaela Frincu, et al. 2021. "Valorization Potentials of Rapeseed Meal in a Biorefinery Perspective: Focus on Nutritional and Bioactive Components." *Molecules* 26 (22). https://doi.org/10.3390/molecules26226787.

Lestari, Dianika, Ednah Zvinavashe, and Johan P.M. Sanders. 2015. "Economic Valuation of Potential Products from Jatropha Seed in Five Selected Countries: Zimbabwe, Tanzania, Mali, Indonesia, and The Netherlands." *Biomass and Bioenergy* 74 (March): 84–91. https://doi.org/10.1016/j.biombioe.2014.12.011.

Letti, Luiz Alberto Junior, Adenise Lorenci Woiciechowski, Adriane Bianchi Pedroni Medeiros, Júlio César de Carvalho, Luciana Porto de Souza Vandenberghe, Susan Grace Karp, Luis Alberto Zevallos Torres, et al. 2021. "Valorization of Solid and Liquid Wastes from Palm Oil Industry." In *Waste Biorefinery: Value Addition through Resource Utilization*, edited by Thallada Bhaskar, Ashok Pandey, Sunita Varjani, and Eldon R. Rene, 235–65. Elsevier. https://doi.org/10.1016/B978-0-12-821879-2.00009-0.

Lhamo, Pema, Shishir Kumar Behera, and Biswanath Mahanty. 2021. "Process Optimization, Metabolic Engineering Interventions and Commercialization of Microbial Polyhydroxyalkanoates Production – A State-of-the Art Review." *Biotechnology Journal* 16 (9): 2100136. https://doi.org/10.1002/biot.202100136.

Liu, Fang, Xiang Guo, Yuan Wang, Guanyi Chen, and Li'an Hou. 2021. "Process Simulation and Economic and Environmental Evaluation of a Corncob-Based Biorefinery System." *Journal of Cleaner Production* 329 (December): 129707. https://doi.org/10.1016/j.jclepro.2021.129707.

Liu, Jingyun, Senqing Fan, Ke Bai, and Zeyi Xiao. 2021. "Combining Acetone-Butanol-Ethanol Production and Methyl Orange Decolorization in Wastewater by Fermentation with Solid Food Waste as Substrate." *Renewable Energy* 179 (December): 2246–55. https://doi.org/10.1016/J.RENENE.2021.08.055.

Liu, Li, Yancui Wang, Na Wang, Xiaomiao Chen, Baoguo Li, Jiping Shi, and Xiang Li. 2021. "Process Optimization of Acetone-Butanol-Ethanol Fermentation Integrated with Pervaporation for Enhanced Butanol Production." *Biochemical Engineering Journal* 173 (September): 108070. https://doi.org/10.1016/J.BEJ.2021.108070.

Liu, Xiaoyan, Fenfen Zhu, Rongyan Zhang, Luyao Zhao, and Juanjuan Qi. 2021. "Recent Progress on Biodiesel Production from Municipal Sewage Sludge." *Renewable and Sustainable Energy Reviews* 135 (August 2020). https://doi.org/10.1016/j.rser.2020.110260.

Liu, Yang, Yizheng Lyu, Jinping Tian, Jialing Zhao, Ning Ye, Yongming Zhang, and Lujun Chen. 2021. "Review of Waste Biorefinery Development towards a Circular Economy: From the Perspective of a Life Cycle Assessment." *Renewable and Sustainable Energy Reviews* 139 (April): 110716. https://doi.org/10.1016/j.rser.2021.110716.

Lu, Yudan, Xiaomei Zheng, Yu Wang, Lihui Zhang, Lixian Wang, Yu Lei, Tongcun Zhang, Ping Zheng, and Jibin Sun. 2022. "Evaluation of Aspergillus Niger Six Constitutive Strong Promoters by Fluorescent-Auxotrophic Selection Coupled with Flow Cytometry: A Case for Citric Acid Production." *Journal of Fungi* 8 (6): 568. https://doi.org/10.3390/jof8060568.

Maddikunta, Praveen Kumar Reddy, Quoc-Viet Pham, Prabadevi B, N Deepa, Kapal Dev, Thippa Reddy Gadekallu, Rukhsana Ruby, and Madhusanka Liyanage. 2022. "Industry 5.0: A Survey on Enabling Technologies and Potential Applications." *Journal of Industrial Information Integration* 26 (March): 100257. https://doi.org/10.1016/j.jii.2021.100257.

Magalhães Júnior, Antonio Irineudo, Carlos Ricardo Soccol, Marcela Candido Camara, Denisse Tatiana Molina Aulestia, Luciana Porto de Souza Vandenberghe, and Júlio Cesar de Carvalho. 2021. "Challenges in the Production of Second-Generation Organic Acids (Potential Monomers for Application in Biopolymers)." *Biomass and Bioenergy* 149 (June): 106092. https://doi.org/10.1016/j.biombioe.2021.106092.

Mahmud, Nazira, and Kurt A. Rosentrater. 2019. "Life-Cycle Assessment (LCA) of Different Pretreatment and Product Separation Technologies for Butanol Bioprocessing from Oil Palm Frond." *Energies* 13 (1): 155. https://doi.org/10.3390/en13010155.

Malik, Sana, Ayesha Shahid, Michael J. Betenbaugh, Chen Guang Liu, and Muhammad Aamer Mehmood. 2022. "A Novel Wastewater-Derived Cascading Algal Biorefinery Route for Complete Valorization of the Biomass to Biodiesel and Value-Added Bioproducts." *Energy Conversion and Management* 256 (February). https://doi.org/10.1016/j.enconman.2022.115360.

Manaf, Intan Shafinaz Abd, Nurul Hajar Embong, Siti Norhafiza Mohd Khazaai, Mohd Hasbi Ab Rahim, Mashitah M. Yusoff, Keat Teong Lee, and Gaanty Pragas Maniam. 2019. "A Review for Key Challenges of the Development of Biodiesel Industry."

Energy Conversion and Management 185 (April): 508–17. https://doi.org/10.1016/J.ENCONMAN.2019.02.019.

Markerts and Markets. 2017. "*Organic Acids Market - Global Forecast to 2022.*" https://www.marketsandmarkets.com/Market-Reports/organic-acid-market-30190158.html#:~:text=%5B174 Pages%5D The global organic,5.5%25 during the forecast period.

Market Data Forecast. 2022. "Biodiesel Market Size, Share & Trends - 2022 to 2027." Biodiesel Market - By Feedstock Type (Vegetable Oils, Animal Fats, Brown Grease/ Trap Grease), Application (Fuel, Power Generation) & By Region (North America, Latin America, Europe, Asia Pacific, Middle East & Africa) - *Global Industry Analysis*, Size, Sh. 2022.

Market Data Forest. 2021. "*Global Xylooligosaccharides Market Research Reports 2022.*" https://www.precisionreports.co/global-xylooligosaccharides-xos-market-21186875.

Mastropetros, Savvas Giannis, Konstantinos Pispas, Dimitris Zagklis, Sameh S. Ali, and Michael Kornaros. 2022. "Biopolymers Production from Microalgae and Cyanobacteria Cultivated in Wastewater: Recent Advances." *Biotechnology Advances* 60 (November): 107999. https://doi.org/10.1016/j.biotechadv.2022.107999.

Medina, Jesus David Coral, Fabio Bolaños Alomia, Antonio Irineudo Magalhaes, Julio Cesar de Carvalho, Adenise Lorenci Woiciechowsky, and Carlos Ricardo Soccol. 2021. "Simulation of Different Biorefinery Configuration Including Environmental, Technical and Economic Assay Using Sugarcane Bagasse." *Journal of Cleaner Production* 316 (September): 128162. https://doi.org/10.1016/j.jclepro.2021.128162.

Meegoda, Jay N., Brian Li, Kush Patel, and Lily B. Wang. 2018. "A Review of the Processes, Parameters, and Optimization of Anaerobic Digestion." *International Journal of Environmental Research and Public Health*. https://doi.org/10.3390/ijerph15102224.

Meramo-Hurtado, Samir I., Ángel González-Delgado, Lars Rehmann, Edgar Quinones-Bolanos, and Mehrab Mehvar. 2021. "Comparative Analysis of Biorefinery Designs Based on Acetone-Butanol-Ethanol Fermentation under Exergetic, Techno-Economic, and Sensitivity Analyses towards a Sustainability Perspective." *Journal of Cleaner Production* 298 (May): 126761. https://doi.org/10.1016/J.JCLEPRO.2021.126761.

Merzari, Fabio, Michela Langone, Gianni Andreottola, and Luca Fiori. 2019. Methane Production from Process Water of Sewage Sludge Hydrothermal Carbonization. A Review. Valorising Sludge through Hydrothermal Carbonization. *Critical Reviews in Environmental Science and Technology*. Vol. 49. https://doi.org/10.1080/10643389.2018.1561104.

Mibielli, Guilherme Martinez, Ana Paula Fagundes, Letícia Renata Bohn, Matheus Cavali, Auriane Bueno, João Paulo Bender, and José Vladimir Oliveira. 2020. "Enzymatic Production of Methyl Esters from Low-Cost Feedstocks." *Biocatalysis and Agricultural Biotechnology* 24 (February). https://doi.org/10.1016/j.bcab.2020.101558.

Milessi, Thais S., Felipe A.S. Corradini, João V.M. Marçal, Thaís O. Baldez, Willian Kopp, Roberto C. Giordano, and Raquel L.C. Giordano. 2021. "Xylooligosaccha-rides Production Chain in Sugarcane Biorefineries: From the Selection of Pretreatment

Conditions to the Evaluation of Nutritional Properties." *Industrial Crops and Products* 172 (November): 114056. https://doi.org/10.1016/j.indcrop.2021.114056.

Miranda, Nahieh Toscano, Ingrid Lopes Motta, Rubens Maciel Filho, and Maria Regina Wolf Maciel. 2021. "Sugarcane Bagasse Pyrolysis: A Review of Operating Conditions and Products Properties." *Renewable and Sustainable Energy Reviews* 149 (July). https://doi.org/10.1016/j.rser.2021.111394.

Mohammadi, Mitra, Mohd Ali Hassan, Lai-Yee Phang, Yoshihito Shirai, Hasfalina Che Man, and Hidayah Ariffin. 2012. "Intracellular Polyhydroxyalkanoates Recovery by Cleaner Halogen-Free Methods towards Zero Emission in the Palm Oil Mill." *Journal of Cleaner Production* 37 (December): 353–60. https://doi.org/10.1016/j.jclepro.2012.07.038.

Morales, Merten, Meriç Ataman, Sara Badr, Sven Linster, Ioannis Kourlimpinis, Stavros Papadokonstantakis, Vassily Hatzimanikatis, and Konrad Hungerbühler. 2016. "Sustainability Assessment of Succinic Acid Production Technologies from Biomass Using Metabolic Engineering." *Energy & Environmental Science* 9 (9): 2794–2805. https://doi.org/10.1039/C6EE00634E.

Nahavandi, Saeid. 2019. "Industry 5.0—A Human-Centric Solution." *Sustainability* 11 (16): 4371. https://doi.org/10.3390/su11164371.

Nandakumar, Ardra, Jo-Ann Chuah, and Kumar Sudesh. 2021. "Bioplastics: A Boon or Bane?" *Renewable and Sustainable Energy Reviews* 147 (September): 111237. https://doi.org/10.1016/j.rser.2021.111237.

Narancic, Tanja, Federico Cerrone, Niall Beagan, and Kevin E. O'Connor. 2020. "Recent Advances in Bioplastics: Application and Biodegradation." *Polymers* 12 (4): 920. https://doi.org/10.3390/polym12040920.

Nayl, A A, A. I. Abd-Elhamid, Ashraf A. Aly, and Stefan Brase. 2022. "Recent Progress in the Applications of Silica-Based Nanoparticles." *RSC Advances* 12 (22): 13706–26. https://doi.org/10.1039/d2ra01587k.

Nazarko, Joanicjusz, Ewa Chodakowska, and Łukasz Nazarko. 2022. "Evaluating the Transition of the European Union Member States towards a Circular Economy." *Energies* 15 (11): 3924. https://doi.org/10.3390/en15113924.

Neyt, Nicole C., and Darren L. Riley. 2021. "Application of Reactor Engineering Concepts in Continuous Flow Chemistry: A Review." *Reaction Chemistry & Engineering* 6: 1295–1326. https://doi.org/10.1039/D1RE00004G.

Nitkiewicz, Tomasz, Magdalena Wojnarowska, Mariusz Sołtysik, Adam Kaczmarski, Tomasz Witko, Carlo Ingrao, and Maciej Guzik. 2020. "How Sustainable Are Biopolymers? Findings from a Life Cycle Assessment of Polyhydroxyalkanoate Production from Rapeseed-Oil Derivatives." *Science of the Total Environment* 749. https://doi.org/10.1016/j.scitotenv.2020.141279.

Odetoye, T. E., J. O. Agu, and E. O. Ajala. 2021. "Biodiesel Production from Poultry Wastes: Waste Chicken Fat and Eggshell." *Journal of Environmental Chemical Engineering* 9 (4): 105654. https://doi.org/10.1016/J.JECE.2021.105654.

Oh, Hurok, Young-Jung Wee, Jong-Sun Yun, Seung Ho Han, Sangwon Jung, and Hwa-Won Ryu. 2005. "Lactic Acid Production from Agricultural Resources as Cheap Raw Materials." *Bioresource Technology* 96 (13): 1492–98. https://doi.org/10.1016/j.biortech.2004.11.020.

Okano, Kenji, Shinji Hama, Maki Kihara, Hideo Noda, Tsutomu Tanaka, and Akihiko Kondo. 2017. "Production of Optically Pure D-Lactic Acid from Brown Rice Using Metabolically Engineered Lactobacillus Plantarum." *Applied Microbiology and Biotechnology* 101 (5): 1869–75. https://doi.org/10.1007/s00253-016-7976-8.

Onday, Ozgur. 2019. "Japan's Society 5.0: Going Beyond Industry 4.0." *Business and Economics Journal* 10 (2): 1000389. https://doi.org/10.4172/2151-6219.1000389.

ONU. 2015. "*United Nations Organization. The Millennium Development Goals Report.*" http://www.agenda2030.com.br/os_ods/.

Österberg, Monika, Mika H. Sipponen, Bruno D. Mattos, and Orlando J. Rojas. 2020. "Spherical Lignin Particles: A Review on Their Sustainability and Applications." *Green Chemistry* 22 (9): 2712–33. https://doi.org/10.1039/D0GC00096E.

Paladino, Ombretta, and Matteo Neviani. 2022. "Sustainable Biodiesel Production by Transesterification of Waste Cooking Oil and Recycling of Wastewater Rich in Glycerol as a Feed to Microalgae." *Sustainability* 14 (1). https://doi.org/https://doi.org/10.3390/su14010273.

Panagiotopoulos, I. A., S. Pasias, R. R. Bakker, T. de Vrije, N. Papayannakos, P. A.M. Claassen, and E. G. Koukios. 2013. "Biodiesel and Biohydrogen Production from Cotton-Seed Cake in a Biorefinery Concept." *Bioresource Technology* 136 (May): 78–86. https://doi.org/10.1016/J.BIORTECH.2013.02.061.

Pinales-Márquez, César D., Rosa M. Rodríguez-Jasso, Rafael G. Araújo, Araceli Loredo-Treviño, Debora Nabarlatz, Beatriz Gullón, and Héctor A. Ruiz. 2021. "Circular Bioeconomy and Integrated Biorefinery in the Production of Xylooligosaccharides from Lignocellulosic Biomass: A Review." *Industrial Crops and Products* 162: 113274. https://doi.org/10.1016/j.indcrop.2021.113274.

Pingkuo, Liu, and Han Xue. 2022. "Comparative Analysis on Similarities and Differences of Hydrogen Energy Development in the World's Top 4 Largest Economies: A Novel Framework." *International Journal of Hydrogen Energy* 47 (16): 9485–9503. https://doi.org/10.1016/J.IJHYDENE.2022.01.038.

Pinto, Ariane S.S., Andrew M. Elias, Felipe F. Furlan, Marcelo P.A. Ribeiro, Roberto C. Giordano, and Cristiane S. Farinas. 2022. "Strategies to Reduce the Negative Impact of Inhibitors in Biorefineries: A Combined Techno-Economic and Life Cycle Assessment." *Journal of Cleaner Production* 345 (April): 131020. https://doi.org/10.1016/j.jclepro.2022.131020.

Polat, Leyla, and Alptekin Erkollar. 2021. "Industry 4.0 vs. Society 5.0." In *Digital Conversion on the Way to Industry 4.0*, 333–45. https://doi.org/10.1007/978-3-030-62784-3_28.

Praptyana, I. R., and Budiyono. 2022. "Biohydrogen Production from Wood Dust Mahogany (Swietenia Mahagony) by Dark Fermentation Using Enterobacter Aerogenes: Effect of Ozone Pretreatment Time and PH." *Materials Today: Proceedings*, March. https://doi.org/10.1016/J.MATPR.2022.02.406.

Preethi, J. Rajesh Banu, Gopalakrishnakumar, Vinay Kumar Tyagi, Amit Kumar Bajhaiya, Poornachandar Gugulothu, and M. Gunasekaran. 2022. "Biohydrogen Production from Waste Activated Sludge through Thermochemical Mechanical Pretreatment." *Bioresource Technology* 358 (August): 127301. https://doi.org/10.1016/J.BIORTECH.2022.127301.

Primmer, Nick. 2021. "*Biogas: Pathways to 2030* - Report - World Biogas Association."

Qiao, Hui, Mingyang Han, Shuiping Ouyang, Zhaojuan Zheng, and Jia Ouyang. 2022. "An Integrated Lignocellulose Biorefinery Process: Two-Step Sequential Treatment with Formic Acid for Efficiently Producing Ethanol and Furfural from Corn Cobs." *Renewable Energy* 191 (May): 775–84. https://doi.org/10.1016/j.renene.2022.04.027.

Reena, Rooben, Raveendran Sindhu, Palanisamy Athiyaman Balakumaran, Ashok Pandey, Mukesh Kumar Awasthi, and Parameswaran Binod. 2022. "Insight into Citric Acid: A Versatile Organic Acid." *Fuel* 327 (November): 125181. https://doi.org/10.1016/j.fuel.2022.125181.

Research and Markets. 2021. "*Biogas Market Size, Share & Trends Analysis* Report By Source (Municipal, Agricultural), By Application (Vehicle Fuel, Electricity), By Region (North America, Europe, APAC, CSA), and Segment Forecasts, 2021-2028."

Research, Precedence. 2022. "*Ethanol Market - Global Market Size, Share, Trends Analysis, Segment Forecasts, Regional Outlook 2021 - 2030.*"

Reshmy, R., Eapen Philip, Aravind Madhavan, Ranjna Sirohi, Arivalagan Pugazhendhi, Parameswaran Binod, Mukesh Kumar Awasthi, Narisetty Vivek, Vinod Kumar, and Raveendran Sindhu. 2022. "Lignocellulose in Future Biorefineries: Strategies for Cost-Effective Production of Biomaterials and Bioenergy." *Bioresource Technology* 344 (January): 126241. https://doi.org/10.1016/j.biortech.2021.126241.

Rodrigues, Caroline Varella, Maurílio Gustavo Nespeca, Isabel Kimiko Sakamoto, José Eduardo de Oliveira, Maria Bernadete Amâncio Varesche, and Sandra Imaculada Maintinguer. 2019. "Bioconversion of Crude Glycerol from Waste Cooking Oils into Hydrogen by Sub-Tropical Mixed and Pure Cultures." *International Journal of Hydrogen Energy* 44 (1): 144–54. https://doi.org/10.1016/J.IJHYDENE.2018.02.174.

Rosenboom, Jan-Georg, Robert Langer, and Giovanni Traverso. 2022. "Bioplastics for a Circular Economy." *Nature Reviews Materials* 7 (2): 117–37. https://doi.org/10.1038/s41578-021-00407-8.

Saadatinavaz, Fateme, Keikhosro Karimi, and Joeri F.M. Denayer. 2021. "Hydrothermal Pretreatment: An Efficient Process for Improvement of Biobutanol, Biohydrogen, and Biogas Production from Orange Waste via a Biorefinery Approach." *Bioresource Technology* 341 (December): 125834. https://doi.org/10.1016/j.biortech.2021.125834.

Sadhukhan, Jhuma, Elias Martinez-Hernandez, Richard J. Murphy, Denny K.S. Ng, Mimi H. Hassim, Kok Siew Ng, Wan Yoke Kin, Ida Fahani Md Jaye, Melissa Y. Leung Pah Hang, and Viknesh Andiappan. 2018. "Role of Bioenergy, Biorefinery and Bioeconomy in Sustainable Development: Strategic Pathways for Malaysia." *Renewable and Sustainable Energy Reviews* 81 (January): 1966–87. https://doi.org/10.1016/j.rser.2017.06.007.

Sanchez, A., I. Valdez-Vazquez, A. Soto, S. Sánchez, and D. Tavarez. 2017. "Lignocellulosic N-Butanol Co-Production in an Advanced Biorefinery Using Mixed Cultures." *Biomass and Bioenergy* 102 (July): 1–12. https://doi.org/10.1016/J.BIOMBIOE.2017.03.023.

Santos, João Vitor dos, Laís Gomes Fregolente, Marcio Justi Laranja, Altair Benedito Moreira, Odair Pastor Ferreira, and Márcia Cristina Bisinoti. 2021. "Hydrothermal

Carbonization of Sugarcane Industry By-Products and Process Water Reuse: Structural, Morphological, and Fuel Properties of Hydrochars." *Biomass Conversion and Biorefinery*. https://doi.org/10.1007/s13399-021-01476-z.

Sayar, Nihat Alpagu, Dilek Kazan, Orkun Pinar, Berna Sarıyar Akbulut, and Ahmet Alp Sayar. 2018. "Retro-Techno-Economic Evaluation of Acetic Acid Production Using Cotton Stalk as Feedstock." *Journal of Material Cycles and Waste Management* 20 (4): 2077–88. https://doi.org/10.1007/s10163-018-0762-1.

Scapini, Thamarys, Maicon S.N. dos Santos, Charline Bonatto, João H.C. Wancura, Jéssica Mulinari, Aline F. Camargo, Natalia Klanovicz, et al. 2021. "Hydrothermal Pretreatment of Lignocellulosic Biomass for Hemicellulose Recovery." *Bioresource Technology*, September, 126033. https://doi.org/10.1016/j.biortech.2021.126033.

Schneider, Willian Daniel Hahn, Aldo José Pinheiro Dillon, and Marli Camassola. 2021. "Lignin Nanoparticles Enter the Scene: A Promising Versatile Green Tool for Multiple Applications." *Biotechnology Advances* 47 (March): 107685. https://doi.org/10.1016/j.biotechadv.2020.107685.

Schultz, Emerson Léo, Daniela Tatiane De Souza, Mônica Caramez, and Triches Damaso. 2014. "The Glycerol Biorefinery : A Purpose for Brazilian Biodiesel Production." *Chemical and Biological Technologies in Agriculture* 1 (7): 1–9.

Selvaraj, Paul Sebastian, Kalaiselvi Periasamy, Kathirvel Suganya, Kavitha Ramadass, Selvamurugan Muthusamy, Poornima Ramesh, Richard Bush, Salom Gnana Thanga Vincent, and Thava Palanisami. 2020. "Novel Resources Recovery from Anaerobic Digestates: Current Trends and Future Perspectives." *Critical Reviews in Environmental Science and Technology* 0 (0): 1–85. https://doi.org/10.1080/10643389.2020.1864957.

Sharma, Shweta, Paras Kumar Jha, and Amit Panwar. 2021. "Production of Bioethanol from Wheat Straw via Optimization of Co-Culture Conditions of Bacillus Licheniformis and Saccharomyces Cerevisiae." *Discover Energy* 1 (1). https://doi.org/10.1007/s43937-021-00004-4.

Sheppard, Phil, Guillermo Garcia-Garcia, Athanasios Angelis-Dimakis, Grant M. Campbell, and Shahin Rahimifard. 2019. "Synergies in the Co-Location of Food Manufacturing and Biorefining." *Food and Bioproducts Processing* 117 (September): 340–59. https://doi.org/10.1016/j.fbp.2019.08.001.

Sherpa, Knawang Chhunji, Debajyoti Kundu, Subhodeep Banerjee, Makarand Madhao Ghangrekar, and Rintu Banerjee. 2022. "An Integrated Biorefinery Approach for Bioethanol Production from Sugarcane Tops." *Journal of Cleaner Production* 352 (August 2021): 131451. https://doi.org/10.1016/j.jclepro.2022.131451.

Silva, Paulo Roberto da, and Eliana M. Alhadeff. 2022. "Biodiesel from Beef Tallow : A Technological Patent Mapping." *Brazilian Journal of Development* 8 (5): 38061–75. https://doi.org/10.34117/bjdv8n5-351.

Silveira, Mariana S., Cláudia P. M. L. Fontes, Alexandre A. Guilherme, Fabiano A. N. Fernandes, and Sueli Rodrigues. 2012. "Cashew Apple Juice as Substrate for Lactic Acid Production." *Food and Bioprocess Technology* 5 (3): 947–53. https://doi.org/10.1007/s11947-010-0382-9.

Singh, Rudra Pratap, H. V. Gangadharappa, and K. Mruthunjaya. 2017. "Phospholipids: Unique Carriers for Drug Delivery Systems." *Journal of Drug Delivery Science and Technology* 39 (June): 166–79. https://doi.org/10.1016/J.JDDST.2017.03.027.

Solarte-Toro, Juan Camilo, and Carlos Ariel Cardona Alzate. 2021. "Biorefineries as the Base for Accomplishing the Sustainable Development Goals (SDGs) and the Transition to Bioeconomy: Technical Aspects, Challenges and Perspectives." *Bioresource Technology* 340 (November): 125626. https://doi.org/10.1016/j.biortech.2021.125626.

Solomou, Konstantina, Mohammad Alyassin, Athanasios Angelis-Dimakis, and Grant M. Campbell. 2022. "Arabinoxylans: A New Class of Food Ingredients Arising from Synergies with Biorefining, and Illustrating the Nature of Biorefinery Engineering." *Food and Bioproducts Processing* 132 (March): 83–98. https://doi.org/10.1016/j.fbp.2021.12.007.

Soltaninejad, Ali, Mohammadhadi Jazini, and Keikhosro Karimi. 2022. "Biorefinery for Efficient Xanthan Gum, Ethanol, and Biogas Production from Potato Crop Residues." *Biomass and Bioenergy* 158 (January): 106354. https://doi.org/10.1016/j.biombioe.2022.106354.

Squinca, Paula, Stanley Bilatto, Alberto C. Badino, and Cristiane S. Farinas. 2020. "Nanocellulose Production in Future Biorefineries: An Integrated Approach Using Tailor-Made Enzymes." *ACS Sustainable Chemistry & Engineering* 8 (5): 2277–86. https://doi.org/10.1021/acssuschemeng.9b06790.

———. 2022. "The Use of Enzymes to Isolate Cellulose Nanomaterials: A Systematic Map Review." *Carbohydrate Polymer Technologies and Applications* 3 (June): 100212. https://doi.org/10.1016/j.carpta.2022.100212.

Steinke, Gleiciéli, Matheus Cavali, João H C Wancura, Jacir Dal, and Magro Wagner. 2021. "Lipase and Phospholipase Combination for Biodiesel Production from Crude Soybean Oil." *Bioenergy Research* 15 (0123456789). https://doi.org/https://doi.org/10.1007/s12155-021-10364-3.

Tanzil, Abid H., Xiao Zhang, Michael Wolcott, Kristin Brandt, Claudio Stöckle, Ganti Murthy, and Manuel Garcia-Perez. 2021. "Evaluation of Dry Corn Ethanol Bio-Refinery Concepts for the Production of Sustainable Aviation Fuel." *Biomass and Bioenergy* 146 (August 2019). https://doi.org/10.1016/j.biombioe.2020.105937.

Teh, Khai Chyi, Jully Tan, and Irene Mei Leng Chew. 2021. "An Integrated Lignocellulosic Biorefinery Design for Nanomaterial and Biochemical Production Using Oil Palm Biomass." *Clean Technologies and Environmental Policy* 23 (10): 2955–73. https://doi.org/10.1007/s10098-021-02215-8.

Toldra-Reig, Fidel, Leticia Mora, and Fidel Toldra. 2020. "Trends in Biodiesel Production from Animal Fat Waste." *Applied Science* 10 (3644): 1–17.

Ubando, Aristotle T., Charles B. Felix, and Wei-Hsin Chen. 2020. "Biorefineries in Circular Bioeconomy: A Comprehensive Review." *Bioresource Technology* 299 (March): 122585. https://doi.org/10.1016/j.biortech.2019.122585.

United Nations. 2015. "*Transforming Our World: The 2030 Agenda for Sustainable Development*." https://www.un.org/ga/search/view_doc.asp?symbol=A/RES/70/1&Lang=E.

Upton, Daniel J., Simon J. McQueen-Mason, and A. Jamie Wood. 2020. "In Silico Evolution of Aspergillus Niger Organic Acid Production Suggests Strategies for Switching Acid Output." *Biotechnology for Biofuels* 13 (1): 27. https://doi.org/10.1186/s13068-020-01678-z.

Vea, Eldbjørg Blikra, Serena Fabbri, Sebastian Spierling, and Mikołaj Owsianiak. 2021. "Inclusion of Multiple Climate Tipping as a New Impact Category in Life Cycle Assessment of Polyhydroxyalkanoate (PHA)-Based Plastics." *Science of The Total Environment* 788 (September): 147544. https://doi.org/10.1016/j.scitotenv.2021.147544.

Velvizhi, G., K. Balakumar, Nagaraj P. Shetti, Ejaz Ahmad, Kamal Kishore Pant, and Tejraj M. Aminabhavi. 2022. "Integrated Biorefinery Processes for Conversion of Lignocellulosic Biomass to Value Added Materials: Paving a Path towards Circular Economy." *Bioresource Technology* 343 (January): 126151. https://doi.org/10.1016/j.biortech.2021.126151.

Wang, Xiao Li, Jin Jie Zhou, Ya Qin Sun, and Zhi Long Xiu. 2019. "Bioconversion of Raw Glycerol from Waste Cooking-Oil-Based Biodiesel Production to 1,3-Propanediol and Lactate by a Microbial Consortium." *Frontiers in Bioengineering and Biotechnology* 7 (FEB): 1–13. https://doi.org/10.3389/fbioe.2019.00014.

Werpy, T, and G Petersen. 2004. "*Top Value Added Chemicals from Biomass: Volume I -- Results of Screening for Potential Candidates from Sugars and Synthesis Gas*." Golden, CO (United States). https://doi.org/10.2172/15008859.

World Bioenergy Association. 2020. "*Global Bioenergy Statistics 2020*." World Bioenergy Association.

Xia, Fei, Huan Liu, Jie Lu, Yanna Lv, Shangru Zhai, Qingda An, Yi Cheng, and Haisong Wang. 2019. "An Integrated Biorefinery Process to Produce Butanol and Pulp from Corn Straw." *Industrial Crops and Products* 140 (November): 111648. https://doi.org/10.1016/J.INDCROP.2019.111648.

Xu, Jikun, Bingchuan Liu, Longsheng Wu, Jingping Hu, Huijie Hou, and Jiakuan Yang. 2019. "A Waste-Minimized Biorefinery Scenario for the Hierarchical Conversion of Agricultural Straw into Prebiotic Xylooligosaccharides, Fermentable Sugars and Lithium-Sulfur Batteries." *Industrial Crops and Products* 129 (March): 269–80. https://doi.org/10.1016/j.indcrop.2018.12.002.

Xu, Li Da, Yang Lu, and Ling Li. 2021. "Embedding Blockchain Technology Into IoT for Security: A Survey." *IEEE Internet of Things Journal* 8 (13): 10452–73. https://doi.org/10.1109/JIOT.2021.3060508.

Xu, Xun, Yuqian Lu, Birgit Vogel-Heuser, and Lihui Wang. 2021. "Industry 4.0 and Industry 5.0—Inception, Conception and Perception." *Journal of Manufacturing Systems* 61 (October): 530–35. https://doi.org/10.1016/j.jmsy.2021.10.006.

Yarali, Abdulrahman. 2021. "Applications of Artificial Intelligence, ML, and DL." In *Intelligent Connectivity*, 279–97. New Jersey: Wiley. https://doi.org/10.1002/9781119685265.ch16.

Yien Ong, Yien, Wen Siang Tan, Mohamad Rosfarizan, Eng Seng Chan, and Beng Ti Tey. 2012. "Isolation and Identification of Lactic Acid Bacteria from Fermented Red Dragon Fruit Juices." *Journal of Food Science* 77 (10): M560–64. https://doi.org/10.1111/j.1750-3841.2012.02894.x.

Younas, Muhammad, Sumeer Shafique, Ainy Hafeez, Fahad Javed, and Fahad Rehman. 2022. "An Overview of Hydrogen Production: Current Status, Potential, and Challenges." *Fuel* 316 (May): 123317. https://doi.org/10.1016/J.FUEL.2022.123317.

Yu, Osbert, and Kwang Ho Kim. 2020. "Lignin to Materials: A Focused Review on Recent Novel Lignin Applications." *Applied Sciences (Switzerland)* 10 (13). https://doi.org/10.3390/app10134626.

Zeng, An-Ping. 2019. "New Bioproduction Systems for Chemicals and Fuels: Needs and New Development." *Biotechnology Advances* 37 (4): 508–18. https://doi.org/10.1016/j.biotechadv.2019.01.003.

Zhang, Changwei, Zhihao Si, Lihe Zhang, Guozhen Li, Jieyi Wen, Changsheng Su, Yilu Wu, Xu Zhang, Di Cai, and Peiyong Qin. 2022. "Reusing the Acetone-Butanol-Ethanol Separated Broth as the Lignocellulose Pretreatment Liquor for Fresh Corn Stalk Biorefinery." *Renewable Energy* 191 (May): 807–18. https://doi.org/10.1016/J.RENENE.2022.04.058.

Zhang, Hongsen, Gang Liu, Jian Zhang, and Jie Bao. 2016. "Fermentative Production of High Titer Gluconic and Xylonic Acids from Corn Stover Feedstock by Gluconobacter Oxydans and Techno-Economic Analysis." *Bioresource Technology* 219 (November): 123–31. https://doi.org/10.1016/j.biortech.2016.07.068.

Zhang, Wei, Juan Pablo Alvarez-Gaitan, Wafa Dastyar, Christopher P. Saint, Ming Zhao, and Michael D. Short. 2018. "Value-Added Products Derived from Waste Activated Sludge: A Biorefinery Perspective." *Water (Switzerland)* 10 (5): 1–20. https://doi.org/10.3390/w10050545.

Zhu, J. Y., Umesh P. Agarwal, Peter N. Ciesielski, Michael E. Himmel, Runan Gao, Yulin Deng, Maria Morits, and Monika Österberg. 2021. "Towards Sustainable Production and Utilization of Plant-Biomass-Based Nanomaterials: A Review and Analysis of Recent Developments." *Biotechnology for Biofuels* 14 (1): 114. https://doi.org/10.1186/s13068-021-01963-5.

Zhu, Jiubin, Wenlu Song, Xue Chen, and Shaoni Sun. 2022. "Integrated Process to Produce Biohydrogen from Wheat Straw by Enzymatic Saccharification and Dark Fermentation." *International Journal of Hydrogen Energy*, May. https://doi.org/10.1016/J.IJHYDENE.2022.05.056.

Chapter 2

Challenges for Biorefineries in Latin America: A Prospective Analysis

María J. Lozano[1,*], Nayda P. Arias Duque[1], Jairo Salcedo Mendoza[1] and Valentina Aristizábal-Marulanda[1,2,*]

[1]Facultad de Ingeniería, Grupo Procesos Agroindustriales y Desarrollo Sostenible (PADES), Universidad de Sucre, Sincelejo, Colombia
[2]Facultad de Tecnologías, Escuela de Tecnología Química, Grupo Desarrollo de Procesos Químicos, Universidad Tecnológica de Pereira, Pereira, Colombia

Abstract

Biorefineries play an essential role in sustainability and the biobased economy. Some of the challenge related to policy regulation, new process, and raw materials needs structural change in the value chain, such as logistics, discovering new raw materials, and overcoming troubles that arose from pandemic SARS Cov 2. It was evident the need to have an auto sustainable economy, the active participation of quadruple helix efforts to drive new developments and reach new milestones in technological maturity.

Government efforts have facilitated the legal framework for the operation of Research Centers with social projection and have allowed, for example, in Costa Rica, to join the developments of nanotechnology to one of the lines of biorefineries where pineapple waste has been taken advantage of it. On the other hand, advances in this sense in Uruguay and Chile have allowed substantial advances in using residues from the meat industry and the production of polyhydroxyalkanoates, respectively.

* Corresponding Author's Emails: valentina.aristizabal2@utp.edu.co; valentina.aristizabal@unisucre.edu.co.

In: The Future of Biorefineries
Editor: Waldemar Nyström
ISBN: 979-8-88697-524-6

Undoubtedly, Brazil is experiencing an important boom in the Latin American economy given its long history in the use of agricultural residues and in obtaining biofuels. Colombia is positioned as an important international actor due to its contributions, the adoption of the 2030 bioeconomy agenda, and the governmental impulse through the 2030 bioeconomy mission for Colombia, which requires the cohesion of the actors from the different industrial sectors, academia, the state, and civil society. In this sense, initiatives developed in post-conflict areas in Colombia for the reconstruction of the social fabric and to provide new energy sources and biobased products from organic agricultural and agro-industrial residues in the Colombian Caribbean are also presented. With the aim of performing a prospective analysis of the state of biorefineries in Latin America, this work was divided into the following section.

1. Bioeconomy and Biorefinery Concept

Georgescu Roegen used the term bioeconomy to refer to a radical ecological perspective on economics he developed in the 1970s and 1980 (Vivien et al. 2019b). It is intended to continually remind us of the biological origin of the economic process and thus highlight the problem of humanity's existence with a limited amount of accessible, unequally located, and unequally appropriated resources (Georgescu-Roegen 1975).

In 2005, the European Union stated that the knowledge-based bioeconomy (KBBE) could be defined as *"the transformation of life science knowledge into new, sustainable, eco-efficient and competitive products"* (European Commission’s Research DG . 2005).

In 2009, the term bioeconomy refers to the set of economic activities related to the invention, development, production, and use of biological products and processes (OCDE 2009). The term evolved in 2012 as "*production of renewable* biological *resources and the conversion of these resources and waste streams into value-added products such as food, feed, biobased products, and bioenergy"* (Union 2012).

The German Federal Government perceives the bioeconomy as *“the production, exploitation, and use of biological resources, processes, and systems to provide products, processes, and services across all economic sectors within the framework of a future-oriented economy.”* (German Federal Goverment. 2021).

The concepts of bioeconomy have advanced over time, resulting in different definitions as new technologies are introduced. Initially, there was talk of Type I bioeconomy, a term that resulted from the need to create a concept that grouped the entire economy of renewable resources, as these resources were under increasing pressure and the first attempts at regulation were being made, particularly concerning fisheries (Vivien et al. 2019a). Subsequently, in 2005 the term type II bioeconomy emerged and was based on the knowledge and adoption of biotechnology components to found the initiative and later became the pillar of the bioeconomy strategy in 2012 (Patermann and Aguilar 2018). It is challenging to accurately define these concepts, as they are relatively new and still under construction. Vivien et al. (Vivien et al. 2019a) establish in their review three types of bioeconomies that were mentioned above. However, the one that relates to the term third generation biorefineries, as shown in Figure 1, has arisen due to the need to cover all the processes of transformation of biomass from various sources in plants with different levels of technological scale.

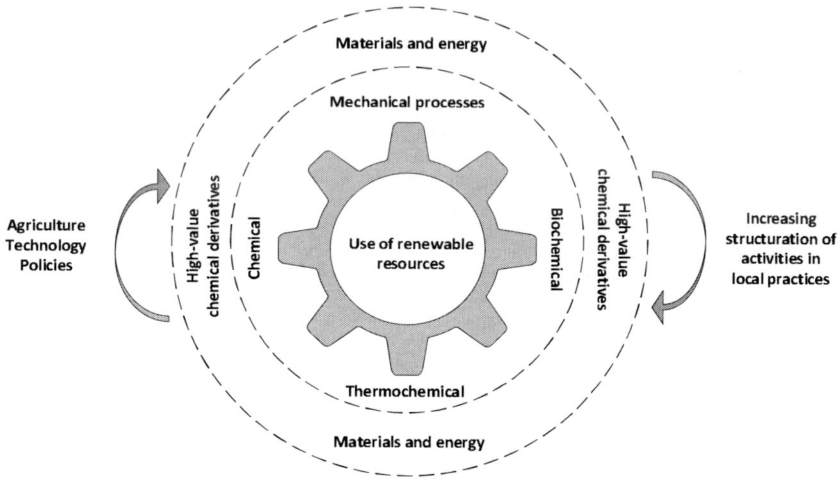

Figure 1. Type III bioeconomy, from (Vivien et al. 2019a).

A correct meshing of biochemical, thermochemical, chemical, and mechanical processes applied to renewable resources should lead to obtaining high-value products through the integration of technological, agricultural, and political aspects. That allow the use of local activity practices based on economic and environmental sustainability. Although this perspective, according to Vivien et al. (Vivien et al. 2019a), does not consider the possibility that these exploitations may face limitations between available and accessible resources.

The Economic Council for Latin America (CEPAL) defines a bioeconomy as the *economy based on the consumption and production of goods and services derived from the direct use and sustainable transformation of biological resources, including biogenic waste generated in the processes of transformation, production, and consumption, Taking advantage of the knowledge of the biological processes and principles, the technologies applicable to the knowledge and transformation of biological resources and the emulation of biological processes and principles* (Rodríguez, Mondaini, and Hitschfeld 2017). As it can be observed, the concept depends on the region that it is applicable. Although the CEPAL definition is an umbrella definition, some Latin American countries define it as shown in Table 1.

Table 1. Bioeconomy definition according to the Latin America country

Country	Bioeconomy Definition	Reference
Argentina	Model for the industrialization of biology, which allows for diversifying export patterns for primary products and agroindustrial commodities through the incorporation of high value-added exports and proven sustainability	(IICA. 2020)
Brazil	Set of economic activities based on the use of renewable biological resources (biomass) in place of fossil raw materials for the production of food, feed, materials, chemicals, fuels, and energy for the production and promotion of health, sustainable development, growth, and well-being for society.	(Carbonell et al. 2021; Nunes 2017)
Colombia	Production, use, and conservation of biological resources, including related knowledge, science, technology, and innovation to provide information, products, processes, and services in all economic sectors for the purpose of moving toward a sustainable economy	(Departmento Nacional de Planeación 2018)
Costa Rica		(Bio Costa Rica. 2022)
Ecuador	The development of the bioeconomy is approached as a mechanism for the sustainable use of biodiversity and as a strategy for resilience to climate change.	(Hugo Chavarría and and Adrián Rodríguez (ECLAC). 2019)
Mexico	The bioeconomy comprises the economic activities supported by the use of biological resources. In its broadest definition, it includes fauna, flora, microorganisms, genetic resources, and biomass derived from processes such as organic waste. Likewise, the functions and services of non-tangible ecosystems such as cultural, tourist, or scenic beauty assets are added to these.	(BIOFIN 2021)
Uruguay	A multisectoral process is currently under way to design the Sustainable Bioeconomy Strategy (EBS), as part of the "Uruguay 2050" National Development Strategy, under the responsibility of the Office of Planning and Budget (OPP), which is under the direct authority of the Office of the President.	(Hugo Chavarría and and Adrián Rodríguez (ECLAC). 2019)

An exhaustive review (Vivien et al. 2019b) concludes that there is an imbalance between the social, economic, and environmental dimensions within the bioeconomy policy framework. Therefore, regulations and governmental pressures to govern are needed due to the new means of biomass processing without putting at risk other sectors and to avoid the generation of problems around the use of soil, crops, or the intensification of processes resulting from agro-industries (OCDE 2009).

For the biorefineries concept, authors such as Ubando et al. (Ubando, Felix, and Chen 2020); Ramcilovic-Suominen et al. (Ramcilovic-Suominen and Pülzl 2018) agree that the term has become involved due to the public policies that were created by the new advances around the establishment of energy alternatives from these facilities and their implications in the bioeconomy for the mitigation of the effects of environmental pollution.

According to Commission to the Gothenburg European Council (Commission to the Gothenburg European Council 2005), sustainable development is based on four pillars – economic, social, environmental, and global governance - which must be mutually reinforcing. The economic, social, and environmental consequences of all policies must be examined in a coordinated manner and taken into account at the time of their elaboration and adoption to promote the concept of the bioeconomy (Vivien et al. 2019b).

That is why establishing this type of alternative plants for processing raw materials from various usable sources is increasingly in the spotlight. From the point of view of Zou et al. (Zou et al. 2016), the development and use of new energy resources have already entered the golden age due to the need to combat the effects of climate change by the indiscriminate use of fossil fuels and that therefore the concept of the "new energy era" will arrive sooner than expected. This era is directly related to the term biorefinery, which, within its processes, includes energy generation from usable resources.

It is well known that industries require energy to run their processes, obtained from raw materials of fossil origin as mentioned above, which contribute to the deterioration of the environment, becoming the focus of concern of the various organizations dedicated to the care of the planet (Alexis Michael Bazzanella, Ausfelder, and DECHEMA Gesellschaft für Chemische Technik und Biotechnologie e.V. 2017). Therefore, strategies have begun to be generated where renewable resources are used for the reduction of greenhouse gases, and the use of biomass is one of the few options for this purpose, projecting itself as a critical renewable resource for the fulfillment of global climate objectives (Patermann and Aguilar 2018) 2018). As described in Task 42 of the International Energy Agency (IEA Bioenergy 2009), which mentions the advantages of taking advantage of technological integration for the processing of biomass raw materials and how this is related to 11 other tasks, which are the commitment of the countries under this organization, as shown in Figure 2, and in turn describes the advantages of its implementation, from the economic and environmental point of view.

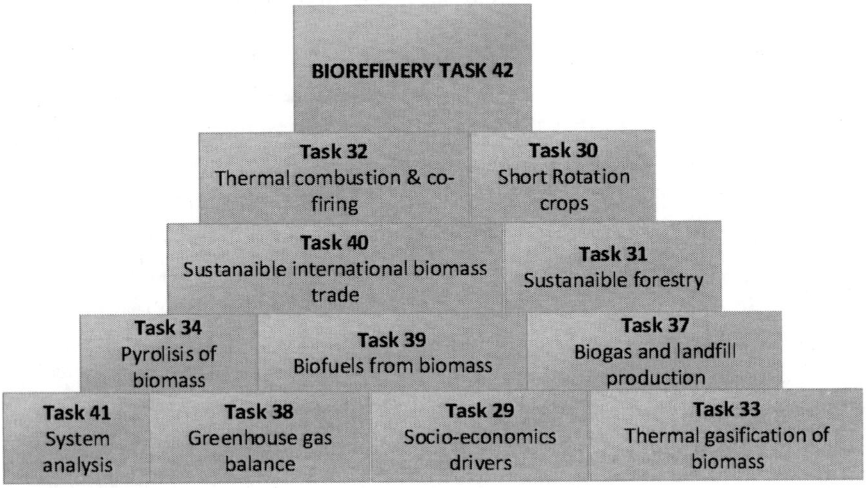

Figure 2. Pillar tasks of IEA bioenergy task 42. Adapted from (IEA Bioenergy 2009).

All these tasks are aimed to be developed within a biorefinery system to obtain different energy products and other bioproducts resulting from the technological integration of several processes, making these systems economically viable. Since the feeding of these plants is versatile, different raw materials can be used, adapting to the environmental conditions and at the same time generating marketable products within the area or beyond it. It should be considered that there are different processing routes for each type of raw material, and the technologies vary accordingly, as shown in Figure 3.

García Prieto (2018) defines this process as one interested in converting biomass into products that satisfy human needs, but this definition establishes that the base product for the transformation is not of fossil origin (García Prieto 2018). Johnson et al., (Johnson and Altman 2014) suggest that this term is even more complex and includes the stages of production, processing, marketing, transportation, and consumption of products of biological origin. From these concepts, biorefinery states that through research, processes can be established that allow the transformation of raw materials from different origins with renewable characteristics, to obtain different platforms that generate other types of products and by-products, whether food or non-food, that have a positive environmental, economic and social impact, that contribute to the improvement of human conditions and comply with the characteristics of global markets.

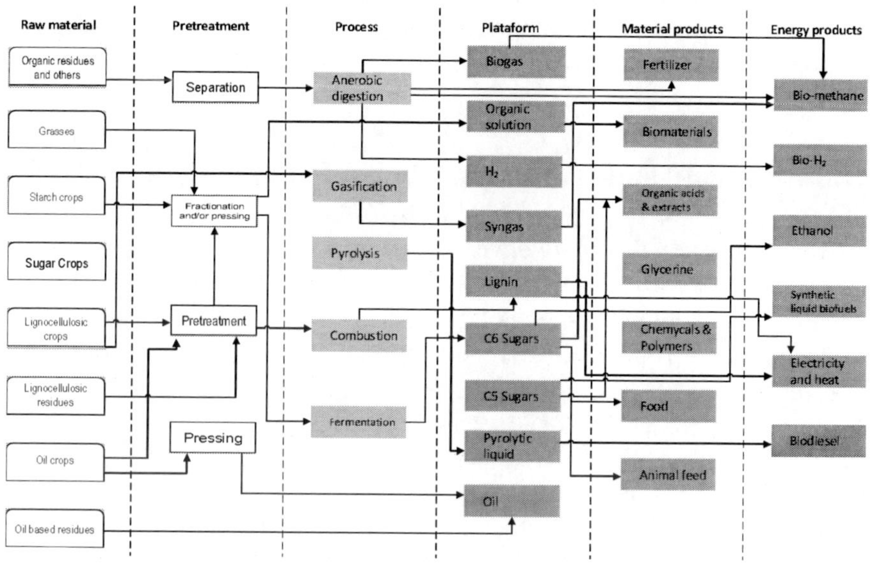

Figure 3. General scheme of a biorefinery with its possible feedstocks, processes, platforms, and final products. Adapted from (Cherubini et al. 2009).

Previously, the concept of biorefineries was defined as units for the transformation of biomass from different sources into a wide variety of energy, chemical, fuel, or other products with different added values (Sacramento-Rivero et al. 2010) Some other authors state that the concept of biorefinery should be defined from the bioeconomy point of view and that it involves physical, chemical or biological processes for the transformation, purification, refining or separation of elements that constitute biological assets of all kingdoms (Conteratto et al. 2021). Kamm and Kamm define biorefinery as "*the intended for final use or for the transformation of biomasses from technological combinations for the interaction between biogenic raw materials with that of intermediate and final products*" (Birgit Kamm and Michael Kamm 2007; Kamm 2014). Currently, this concept is aimed at exploiting and searching for new ways of using available biomass for its application in different types of biorefineries.

The desired product should be considered choosing the most effective route; these routes can be thermochemical (gasification and pyrolysis), biochemical (fermentation and digestion), chemical processes (hydrolysis), and mechanical processes, among others (Aristizábal-Marulanda, Botero-Gutierrez, and Cardona-Alzate 2019; Aristizábal-Marulanda Valentina., Cardona- Alzate, and Martín 2019). Figure 3 shows the production routes and the processes occurring in a biorefinery scheme, which can be classified by platforms, products, feed, and processes.

2. Bioeconomy and Biorefinery Public Policy

As was mentioned before, to develop the bioeconomy in Latin America, it is necessary to provide an adequate policy framework. Different economic and environmental policy has been established in Colombia as follows, The Convention on Biological Diversity (CBD) (ONU. 1992), approved by Colombia through Law 165 of 1994, National Biodiversity Policy of 1996.(Ministerio de Ambiente y Desarrollo Sostenible. 1996), the National Policy for the Comprehensive Management of Biodiversity and its Ecosystem Services PNGIBSE (Ministerio de Ambiente y Desarrollo Sostenible. 2012), CONPES 2834 – Forest Policy (CONPES 2834. 1996), CONPES 3510 - Policy guidelines to promote the sustainable production of biofuels in Colombia (CONPES 3510. 2008), CONPES 3527 - National Competitiveness and Productivity Policy (CONPES 3527. 2008), CONPES 3533 - Basis for an action plan to adapt the intellectual property system to national competitiveness and productivity 2008-2010 (CONPES 3533. 2008), CONPES 3582 - National Policy for the Promotion of Research and Innovation (CONPES 3582. 2009), CONPES 3678 – Productive Transformation Policy: A sectoral development model for Colombia (CONPES 3678. 2010), CONPES 3680 - Guidelines for the consolidation of the System National Protected Areas (CONPES 3678. 2010)), CONPES 3697 - Policy for the commercial development of biotechnology based on the sustainable use of biodiversity (CONPES 3697. 2011), CONPES 3700 – Institutional strategy for the articulation of policies and actions on climate change in Colombia (CONPES 3700. 2011), CONPES 3834 - Policy guidelines to stimulate investment private in science, technology and innovation through deductions taxes (CONPES 3834. 2015), CONPES 3866 – National Productive Development Policy (CONPES 3866. 2016), CONPES

3834 CTi - National science, technology and innovation policy, 2015-2025 , CONPES 3874 - National policy for the comprehensive management of solid waste (CONPES 3874. 2016), CONPES 3892 – Policy guideline to stimulate private investment in science, technology and innovation through deductions taxes (CONPES 3892. 2017)(Conpes3892, 2017), CONPES 3918 of 2018- Strategy for the implementation of the Sustainable Development Goals (SDG) in Colombia (CONPES 3918. 2018), CONPES 3934 Green Growth Policy (CONPES 3934. 2018), CONPES 3975 National Policy for Digital Transformation and Artificial Intelligence (CONPES 3975. 2019), CONPES 3990 Colombia bio-oceanic potency (CONPES 3990. 2020)CONPES 4004 Circular economy in the management of drinking water services and wastewater management (CONPES 4004. 2020), CONPES 4021 National Policy for the Control of Deforestation and the Sustainable Management of Forests (CONPES 4021. 2020), CONPES 4050 Policy for the Consolidation of the National System of Protected Areas SINAP (CONPES 4050. 2021), CONPES 4062 National Intellectual Property Policy (CONPES 4062. 2021), CONPES 4069 National Science Technology and Innovation Policy (CONPES 4069. 2021), CONPES 4075 Energy Transition Policy (CONPES 4075. 2022), CONPES 4085 Internationalization Policy for Regional Productive Development (CONPES 4085. 2022) CONPES 4098 Policy to Promote Agricultural Competitiveness (CONPES 4098. 2022).

These regulatory frameworks allow the Colombian government to implement the Bioeconomy Mission for Colombia; Bioeconomy for a living and diverse power; Towards a knowledge-driven society (Ministerio de Ciencia Tecnología e Innovación. 2020). The Strategic areas and challenges are biodiversity and ecosystemic services, Colombia biosmart, Productive and sustainable agriculture, Biomass and green Chemistry, and Health & Wellness (Ministerio de Ciencia Tecnología e Innovación. 2020). Biorefinery, biofactories, bioenergy and biofuels, biological catalyst, and green chemistry belong to the process of adding value to Biomass and green chemistry strategy. We believe that soon, Colombia will implement small-scale biorefineries or large-scale biorefineries in TRL 8-9. Examples of the biorefinery approach taking into account the agricultural waste and Colombian biodiversity can be observed in the analysis of the routes for biomass processing towards sustainable development in the conceptual design step (Ortiz-Sanchez and Cardona Alzate 2022), production of bioethanol, biogas, syngas, and electricity using coffee-cut stems as raw material (Aristizábal-Marulanda, Solarte-Toro, and Cardona Alzate 2021), The integral use of aromatic plants: prefeasibility comparison of stand-alone and biorefinery processes using thyme (*Thymus vulgaris*) as a base case (Marín-Valencia et al. 2021), The biorefinery approach to orange peel valorization (Ortiz-Sanchez et al. 2021).

On the other hand, Costa Rica is a successful case due to the implementation of the strategies of biorefineries and nanotechnology. Efforts were done to build the "Laboratorio Nacional de Nanotecnología, LANOTEC. It was created with the support of various sectors of Costa Rica, MICIT, CONARE, CENAT and the High Technology Industry, and international institutions such as NASA. LANOTEC began research on October 18, 2004 (José Roberto Vega Baudrit. 2007). LANOTEC strengthened the country, giving it the ability to be the technological leader in the Central American and Caribbean region, with state-of-the-art engineering in the study of advanced materials for research, design, and training in technologies associated with microtechnology, nanotechnology, and materials science (José Roberto Vega Baudrit. 2007). It has allowed expanding the development of knowledge and collaboration with the formation of human resources and scientific research in this area and contributes to developing specific applications for the productive sector in different types of industries such as metallurgical, ceramic materials, polymers, for microbiology, medicine, geophysics and space exploration, among others (José Roberto Vega Baudrit. 2007). In the biorefinery area, this laboratory has studied Banana Waste to Obtain High-Value Biopolymers (Redondo-Gómez et al. 2020), Persea americana biorefinery for value-added products (Mora-Sandí et al. 2021), and pineapple biorefinery (Eixenberger et al. 2022).

The law for Comprehensive waste management (Gobierno de Uruguay. 2019), and the National Plan for Waste Management (Gobierno de Uruguay. 2021) in Uruguay, introduce the framework for the sustainable use of meet wastes (Magdalena et al. 2021). The value-added products obtained from this waste are heparin, keratin hydrolysate (Beatriz Briano Col. 2021), and collagen (Magdalena et al. 2021). Biogas also can be obtained from solid waste (Moreda 2016). Research on anaerobic technology is developed at Universidad de la República (the single public university) in the Chemical Engineering Department (Moreda 2016).

The National strategy of organic waste in Chile proposes that in 2040, Chile can take advantage of two out of every three kilograms of organic waste, putting it at the level of the more developed countries in the matter and advancing decisively towards a circular economy. The strategy proposes a goal to go from 1% to 66% of recovery of organic waste generated at the municipal level by 2040. (Ministerio del Medio Ambiente de Chile. 2020).

In this sense, fruit residues can help obtain Polyhidroxyalkanoates developed from fruit waste (Andler et al. 2021) in Chile. The transformation of the microbial bioplastics, such as the PHA industry, into a bioeconomy with circular product streams can be achieved by utilizing by-products from the fruit industry together with the integration of strategies to reduce energy doing the Scaling up and optimizing the production of an eco-friendly process (Andler et al. 2021).

The bioeconomy strategy in Brazil is driven by the vast development in bioenergy dedicated biorefineries. The National Biofuels Policy (RenovaBio) was launched in December 2017 (Grangeia, Santos, and Lazaro 2022). Introduces a methodology that allocates costs on externalities to achieve environmental targets and gives incentives for the expansion of biofuel production and its use for transportation. This strategy supports the Brazilian Nationally Determined Contribution (NDC)'s commitments under the ratification of the Paris Agreement in 2016. This policy supports national energy security, reducing greenhouse gas (GHG) emissions by 37% below 2005 levels in 2025 and the indicative contribution of 43% reduction below 2005 emission levels in 2030. Mechanisms such as Life Cycle Assessment (LCA), commercialization, and predictability of the fuel market are involved (Grangeia, Santos, and Lazaro 2022). Brazilian soil has important mineral contents and a tropical climate that favors the growth biomass feedstocks used to produce biofuels (Benites-Lazaro et al. 2020). The main crops used for biofuel in Brazil are the sugar cane crops. Brazilian institutionalism explains the almost 500-year-old history of sugar cane (Costa 2020).

3. New and Old Raw Materials

The feedstocks initially explored in biorefinery schemes came directly from lignocellulosic crops (plant biomass), mainly for food use. These, to produce fuels and other types of products for industrial use, are obtained by taking advantage of the polymeric characteristics, as shown in Figure 4. The choice of these raw materials is based mainly on physicochemical analyses that reveal their main characteristics and, therefore, the contents of interest that can be exploited.

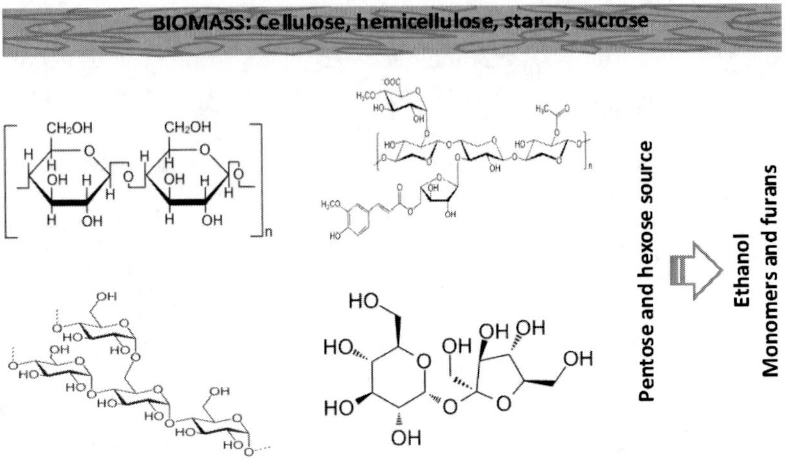

Figure 4. Plant biomass contents.

Such is the case of the sweet sorghum crop, (Nghiem et al. 2013) studied the production of fuel ethanol from this crop, taking advantage of the carbohydrate content of the raw material, which varies between 70.0-90.0%.

Another raw material that aroused great interest in South America, specifically in Brazil, is those resulting from the processing of sugar cane since, due to the availability and large volumes available in this country, which is the world's leading producer of this crop, has processing plants that generate raw materials without any added value, for processing into other products with great industrial potential. It should be noted that, of all its neighbors, this country has the most remarkable technological progress in terms of biorefineries, and most of the established ones are fed from the raw material mentioned above, as shown in Table 2.

Therefore, there is a large generation of waste, such as bagasse and stillage (Caderby et al. 2013). Bagasse could be exploited for its xylan content, which can be used to produce xylose, xylitol, second-generation ethanol, and xylooligosaccharides; the latter is of great value to food companies due to its prebiotic characteristics, which can provide health benefits to consumers (Bragatto, Segato, and Squina 2013). Antioxidants and phenolic compounds were found in the sugarcane stillage in which the fraction rich in phenolic compounds were responsible for 37-59% activity, and 58 to 66% of the global antioxidant activity was associated with the nature of the sugar cane stillage.

As well as these, there are examples around the world of using food crops to feed these plants, which in many sectors has provoked fears about using agricultural land for non-food purposes. Using non-conventional raw materials to produce compounds through biorefinery schemes has become an alternative to minimize this. According to CEPAL (Consejo Económico para América Latina (CEPAL). 2015), the agriculture sector is transforming from a food sector to a complex biomass supply and processing sector.

It has been decided to replace the use of the crops themselves using their residues, plant material that did not imply any use that could generate alternatives in both economic and environmental improvements. Among these crops are rice, one of the most abundant in the world and the main staple of the family food basket in many countries. Another compound of interest from rice residues is biohydrogen. Some studies by Bundhoo et al. (Bundhoo, Mohee, and Hassan 2015) establish that the lignocellulosic material resulting from this source has an energy potential of around 622.6 PJ/year. The rice husk is another residue from this crop (a process-based waste). In 2016, the estimated world production was 203 million tons (Mohan et al. 2018). It has been used as biochar from the slow pyrolysis technique to improve soil fertility and CO_2 retention. Some biochars obtained from rice husks have also demonstrated potential for wastewater purification (Ratan, Kaur, and Adiraju 2018).

Other crops that have emerged as of interest in some Latin American countries due to their significant presence are cassava. It has recently been determined that the residues generated from agricultural and agroindustrial processes have significant lignocellulosic characteristics that allow them to be used for their utilization through techniques such as incineration, pyrolysis, gasification, ethanolic fermentation, anaerobic digestion, among others (Sivamani, Balajii, and Shanmugaprakash 2018). Cassava residues such as peel and bran have been used to produce biochar, biobutanol, bioethanol, biogas, and synthesis gas, with good results at the laboratory scale.

Other studies explore routes for converting these residues into compounds, such as levulinic acid (LA), which imparts carboxylic acid and ketone bifunctionality, serving as a highly versatile chemical platform for the production of various value-added products (Azlan et al. 2022). An emerging raw material for obtaining this compound is bovine manure can achieve yields of up to 338.9 g (LA)/kg by pre-treating the manure with potassium hydroxide (KOH) (Su et al. 2017). The products generated from LA have diverse applications in industries such as food and biofuels from methyl levulinate, polycarbonates, agrochemicals, solvents, fragrances, and plasticizers (Pileidis and Titirici 2016b).

Table 2. Biorefineries established in Latin America

Name	Generation	Country	Production ($Kton/y$)	Feedstock	Refining Technology	Main product	TRL	Status
N. T. University	1	Argentina	0,03	Algae	Algae Oil extraction	Renewable Oil	N.A.	N.A.
Moema	1	Brazil	100	Starch Crops, Sugar crops	Fermentation	Bio-Oil	8	Idle
Amyris	1	Brazil	493	Sugarcane	Fermentation	Farmesene	N.A.	N.A.
PHB Industrial S/A	1	Brazil	0,1	Sugarcane	Fermentation	Polyhydroxyalkanoates (PHA)	N.A.	Operational
Tereos Starch & Sweeteners	1	Brazil	N.A.	Cassava, Corn	Extraction	Starch, Modified Starch, Syrups, Gluten, Fibres, Germs	N.A.	Operational
CTC	1	Brazil	2	Sugarcane	Fermentation	Ethanol	6-7	Operational
Amyris (Piracicaba)	1	Brazil	52	Sugarcane	Fermentation	Farmesene	9	N.A.
Amyris (Campinas)	1	Brazil	83	Sugarcane	Fermentation	Farmesene	9	N.A.
BioVerde	1	Brazil	87	Soy	Transesterification	Biodiesel	N.A.	N.A.
Evonik Industries AG	1	Brazil	200	Corn	Fermentation	Lysine	N.A.	N.A.
Braskem	1	Brazil	200	Sugarcane	Fermentation	Ethylene, Polyethylene	N.A.	N.A.
Bioflex 1	2	Brazil	62	Sugarcane bagasse, Straw	Fermentation	Ethanol	8	Operational
Aracruz Project	2	Brazil	9,6	Forest residues	Circulating fluid bed	Pyrolysis oil	9	Under construction
Dow DuPont	2	Brazil	350	Sugarcane	Fermentation	Ethylene	N.A.	N.A.
Pilot	2	Brazil	0,21	Sugarcane bagasse	Fermentation	Ethanol	4-5	Operational

Name	Generation	Country	Production $(\boldsymbol{Kton/y})$	Feedstock	Refining Technology	Main product	TRL	Status
Raizen Energy	2	Brazil	31,6	Sugar crops residues	Fermentation	Ethanol	8	Operational
JBF industries	Multifeed	Brazil	500	Sugarcane bagasse	Fermentation	Bio-Based Glycol	N.A.	Operational
LS9	Multifeed Stock	Brazil	572	Sugarcane, Corn Syrup, Sweet Sorghum Syrup, Molasses, Glycerin	Bioprocessing	Renewable drop in fuel	9	N.A.
BioAmber	N.C	Brazil	181	N.A	Fermentation	Succinic Acid	9	N.A.
UDT	N.C	Chile	0,08	N.A	Fluid bed	Bio-Oil	1-3	Operational

Adapted from (IEA Bioenergy., Agenzia nazionale per le nuove tecnologie l'energia e lo sviluppo economico sostenibile., and ENEA. 2022).
TRL: 1-3: Research, 4-5: pilot, 6-7: Demonstration, 8: First of a kind commercial 9: commercial, N.A: Not apply.

4. Building Blocks Molecules

In 2004, the National Renewable Laboratory (NREL) and the Pacific Northwest National Laboratory (PNNL) submitted to the Department of Energy of USA (DOE) a list of twelve biobased sugar-based building blocks selected from more than 300 candidates (T. Werpy and G. Petersen 2004). These are 1,4-diacids (succinic, fumaric, and malic), 2,5-furandicarboxylic acid, 3-hydroxypropionic acid, aspartic acid, glucaric acid, glutamic acid, itaconic acid, levulinic acid, 3-hydroxybutyrolactone, glycerol, sorbitol, and xylitol/arabinitol. They were examined by considering the technical complexity of the synthetic pathways and their derivatives' potential markets. Several high-value biobased chemicals can be obtained from these building block molecules (T. Werpy and G. Petersen 2004). The second group of secondary building blocks chemicals also was informed. These secondary chemicals comprise gluconic acid, lactic acid, malonic acid, propionic acid, the triacids citric and aconitic, xylonic acid, acetoin, furfural, levoglucosan, lysine, serine, and threonine (T. Werpy and G. Petersen 2004). NRLE used inclusion criteria such as marketability, properties, yields, previous industry experience, the technical complexity of synthesis routes, and the ability to compete directly with existing petroleum-derived chemicals (T. Werpy and G. Petersen 2004). Although years later, according to the evolution of the selected compounds, a new review was conducted, resulting in the selection of 10 molecules based on nine criteria: the compound or technology has received significant attention in the literature, the compound illustrates a broad technology applicable to multiple products, the technology provides direct substitutes for existing petrochemicals, the technology is applicable to high volume products, a compound exhibits strong potential as a platform, scaleup of the product or technology to pilot, demo, or full scale is underway, the biobased compound is an existing commercial product, prepared at intermediate or commodity levels, the compound may serve as a primary building block of the biorefinery and, commercial production of the compound from renewable carbon is well established (Bozell and Petersen 2010).

When platforms are referred to, they refer to products that link the union of raw materials and marketable end products (T. Werpy and G. Petersen 2004). This term is used in many industries to explain the transformation of raw material into different fractions to obtain other products that may or may not be pure. The production of new compounds, subjected to variable conversion processes, yields different products, as shown in Table 3.

Table 3. Biorefineries and products obtained from different platforms

Biorefinery	Products
Based on synthesis biogas	Syngas, hydrogen, methanol, dimethylether, diesel FT
Pyrolysis-based	Bio-oil, diesel fuel, chemical compounds, oxygenates,
Hydrothermal conversion based	Hydrogen
Fermentation based	C_xH_x, diesel fuel, chemical compounds
Based on oleaginous plants	Bioethanol

IEA Bioenergy 2009; T. Werpy and G. Petersen 2004.

Based on the products obtained, biorefineries can be defined in three classes: biofuels, bioenergy, biochemicals, and materials such as succinic acid, biofuels, and biochemicals. The world economic forum (King 2010) defines biorefinery as a concept analogous to oil refineries. However, its most common classification is by generation: 1st generation (use of biomass), 2nd generation (from lignocellulosic residues to municipal waste), 3rd generation (Biomass from algae), and 4th generation (genetically modified feedstocks with genomically synthesized microorganisms).

All these concepts have led to the study of different compounds regularly known as "platform molecules," obtained from various biomasses rich in hemicellulose and sugars, with great potential to produce industrially important products. However, many of these molecules have been studied and registered worldwide. The following is a description of the top 10 chosen according to the latest report on the subject published by the US Department of Energy. Large chemical companies such as DuPont, BASF, and the Dow Chemical Company actively pursue valorizing lignocellulosic biomass (Takkellapati, Li, and Gonzalez 2018).

4.1. 2,5-Furan Dicarboxylic Acid (FDCA)

2,5-Furan dicarboxylic acid (FDCA), which is a monomer obtained from 5-hydroxy methylfurfuraldehyde (HMF), has shown an excellent capacity for the synthesis of polymers (Wang, Zhang, and Liu 2015). There are several routes for obtaining FDCA from HMF, for example, using microorganisms such as *Aspergillus flavus* APLS-1, resulting in an efficient and environmentally friendly conversion (Rajesh et al. 2019). Also, nanocatalysts based on magnetite and cobalt oxide (Fe_3O_4–CoO_x), can be employed in the catalytic conversion of fructose. The yields of up to 93.1% HMF with lower production costs. Using lignocellulosic biomasses with improved pretreatments to improve glucose yields is the chosen route to reduce the use of fructose because it can incur high economic costs (Santiago and Guirardello 2020).

The catalytic ring-opening reaction coupled with hydrogenation of FDCA was used to obtain the synthetic adipic acid (AA), which is mainly a monomer used in the production of nylon 66 polyurethane, resins, plasticizers, food, and pharmaceutical additives (Vy Tran et al. 2022). FDCA, in combination with 1,4-butanediol, produces Poly(butylene-2,5-furan dicarboxylate) (PBF), with high nucleation densities by differential scanning calorimetry tests (Papageorgiou et al. 2014). Hydroxymethylfurfural (HMF) could become a precursor of furan, which can be used as a green chemical from direct catalytic conversion of bagasse fiber to produce chemicals, biofuels, and solvents of great importance in the industries, especially in the polymer market (Abdulkhani et al. 2021).

This molecule is an important chemical platform obtained by transforming the abundant and cheap hexoses (glucose and fructose), into intermediates such as 5-hydroxymethylfurfural (HMF), which is the result of fructose dehydration and a precursor of FDCA through its oxidation (Chen et al. 2021). The importance of FDCA arises from its chemical structure, as it contains a rigid furan ring and two diacid side chains that can efficiently produce condensation polymers in a manner similar to terephthalic acid (TPA) (Terzopoulou et al. 2020). Being an excellent precursor to produce green polymers such as polyethylene 2,5-furandicarboxylate (PEF), has displaced petroleum derivatives such as TPA, through its production by chemical-catalytic routes, which appears to be the most promising in terms of yield, reaction rate and product purity (Sajid, Zhao, and Liu 2018). Within these routes, thermo catalysis is the most used, as it is a convenient and efficient procedure, although some authors have suggested that electrocatalytic oxidation is a cleaner and safer method (Chen et al. 2021). Thus, through experimentation, techniques such as co-polycondensation with other compounds have been used to improve the tenacity of polyethylene furanoate (PEF), which is the main problem it presents when made from FDCA, resulting in the preservation of its gas barrier properties and improving its capacity as an ecological packaging material (Xie et al. 2019).

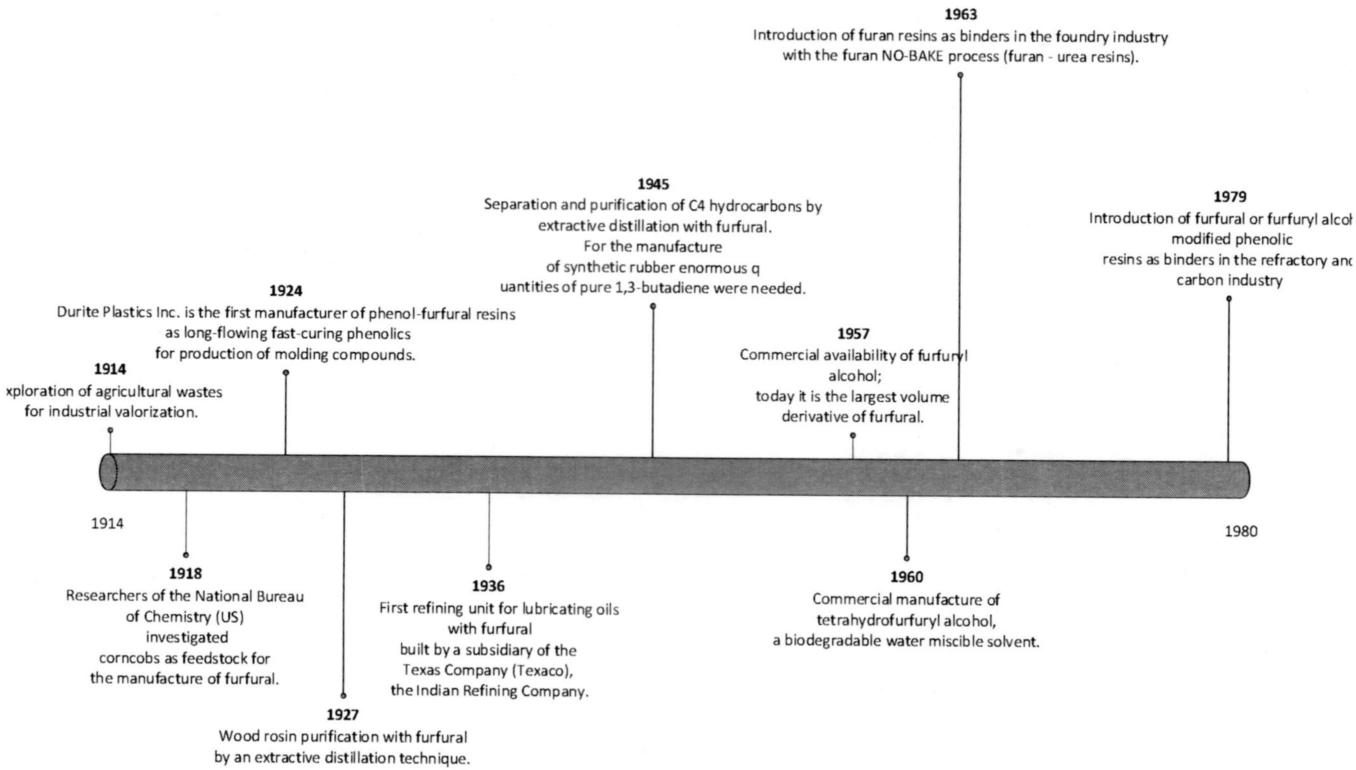

Figure 5. Significant moments of furfural in the industry in its early days. Adapted from (International Furan Chemicals B.V. 2016).

The European Commission (European Commission. 2015), estimated that by 2020 its use within PEF formulations would be around 322000 tons, with other uses ranging from 6000 tons to 80000, operated mainly by key companies such as Avantium, Corbion, Novozymes (European Comission. 2015) Toronto Research Chemicals, and Alfa Aesar, Synvina, Asta Tech, Novamont, Chemsky (Shanghai) International, AVALON Industries, V & V Pharma Industries (Master Watch. 2022).

4.2. Furfural

Furfural was discovered in 1821 by the German chemist Döbereiner, but it was not until 1840 that the correct empirical formula $C_5H_4O_2$ was assigned by Stenhous (Li, Jia, and Wang 2016; International Furan Chemicals B.V. 2016). Furfural is mainly produced by hydrolysis and dehydration of xylan, which exists in large quantities in hemicellulose (Li, Jia, and Wang 2016). The conversion of xylose/xylan into furfural is a well-explored process that was first industrialized in 1921 by the Quaker Oats Company (Brownlee and Miner 1948). The most significant advances in its industrial production are shown in Figure 5.

Furfural is a versatile block molecule that can be obtained from the hemicellulose fraction through acid hydrolysis, recovery, and finally purification through a distillation system (Yong et al. 2022). One of these versatilities is the production of additives such as 2-methyl furan, 2-methyl tetrahydrofuran, and 2-methyl tetrahydrofuran, alkyl levulinates, ethylfurfuryl ether, ethyl tetrahydrofurfuryl ether, tetrahydrofurans, mainly used in fuel blends, generating positive impacts by replacing petrochemical compounds with those obtained from lignocellulosic biomass, which is highly available and generates lower environmental impacts (Ahmad et al. 2022). On the other hand, other types of waste, such as paper, with high ash and lignin content, have been used for furfural production by H_2SO_4 and $FeCl_3$ (Dutta et al. 2022).

4.3. Hydroxypropionic Acid

Hydroxypropionic acid (3-HPA) is a component obtained from biomass containing C5 and C6 sugar. It is a crucial component in producing high-value products, such as acrylic monomers, 1,3-propanediol (PDO), propene, malonic acid, β-propiolactone, and 3-hydroxypropionic esters (Isikgor and

Becer 2015). This block molecule, as mentioned above, can be obtained from biomass as an alternative method to fossil raw materials with better-operating conditions, lower environmental risks, and low carbon dioxide emissions (Mika, Cséfalvay, and Németh 2018). It has been reported that it can be produced from various intermediates, such as glycerol, malonyl-CoA, and β-alanine, as well as by microorganisms such as *Escherichia coli*, yeasts, and cyanobacteria (Liu et al. 2017), by simple bio-based routes based on glycerol utilization. It requires only two enzymatic activities: glycerol dehydratase and aldehyde dehydrogenase (Becker et al. 2015). Currently, most companies in the chemical industry use a dehydrogenation process to produce acrylic acid from 3-HPA, which is used as a base to produce copolymer plastics, adhesives, inks, and other essential products for the industries (Choi et al. 2015).

4.4. Lactic Acid

Lactic acid (LA), also known as 2-hydroxypropanoic acid, is abundant in nature. Scheele discovered it in 1780 from sour milk and determined it was made up of two optical enantiomers (Koutinas et al. 2014). Since its discovery, applications in the cosmetic, pharmaceutical, and food industries have positioned it as a high-value product, which has also granted it Generally Recognized As Safe recognition (GRAS), being identified as safe by the U.S. Food and Drug Administration (Castillo Martinez et al. 2013).

This product occurs primarily by fermentation, which is rated with a yield of up to 90% calcium lactate based on fed glucose, which is neutralized to produce pure lactic acid (Bozell and Petersen 2010). Currently, in the search for more ecological alternatives, other products have been used for the generation of this molecule (Khoshnevisan et al. 2020), where the use of *Lactobacillus delbrueckii* DSM 20074 for the commercial production of lactic acid from municipal waste is evidenced, demonstrating its production from biomass presents ecological solutions compared to the equivalent products that will replace. Thus, one of the most promising applications of this monomer is the replacement of petroleum-based monomers for the synthesis of polylactic acid (Bustamante et al. 2019), which is manufactured by sequential chemical reactions of lactide formation followed by polymerization (Choi et al. 2015) and is mainly used in the bioplastic products industry, representing 11.4% of total production in the world. However, it is estimated to increase by 25% by 2025 (Bustamante et al. 2019). The main challenges in producing lactic acid are the high costs of the starchy biomass used for fermentation, which has led to the use of other types of lignocellulosic biomass. However, the production processes have determined that saccharification is necessary to obtain better results, thus increasing the processing costs (Mazzoli 2020). Studies have suggested that using macroalgae as substrate in biorefinery platforms holds promise for sustainable LA production due to their rapid growth rate and wide range of polysaccharide content (Chung et al. 2021).

4.5. Ethanol

Ethanol was initially not included in the list of DOE (T. Werpy and G. Petersen 2004) because, at that time, its high production volumes classified it as a super product, but due to technological developments in recent years, it was not included in the list of Bozell (Bozell and Petersen 2010). It has become a vital product due to low procurement costs and pretreatment and detoxification technologies, such as combined pretreatment with simultaneous saccharification and fermentation for biomasses from various sources. However, cost-effective methods still need to provide maximum sugar yields with minimum inhibitory compounds, energy, and chemical consumption (Bhatia et al. 2020).

The production of ethanol from lignocellulosic raw materials has become an alternative to replace many petrochemical compounds to obtain ethylene. Recently, countries such as Brazil have studied the use of sweet potatoes due to their high rate of waste in the country. It has been demonstrated that a biorefinery model for obtaining ethanol from ethanol is sustainable and contributes to the development of the agri-food sector by providing new business and consequent job creation (Weber, Trierweiler, and Trierweiler 2020).

There is much information about this platform molecule. However, currently, the main challenge of its production lies in the ability to achieve a positive energy balance and high economic viability with minimal environmental impacts, i.e., energy or product production significantly exceeds energy and cash input (Vu et al. 2020), something that most companies desire, but ethanol is close to achieving it.

4.6. Four Carbon 1,4-Diacids (Succinic, Fumaric and Malic)

The Four Carbon 1,4-Diacids family comprises three elements of great industrial interest. Among these, succinic acid, also known as amber acid or butanedioic acid, is a dicarboxylic acid with the molecular formula $C_4H_6O_4$ (Borges and Pereira Jr. 2011). This block molecule has multiple applications in the industry. However, a fascinating one is a combination with the structures of aliphatic polyamides to create more hydrophilic chains that give these compounds biodegradability characteristics, to generate less environmental impact in its use (Rorrer et al. 2017).

In Brazil, research has been carried out to obtain succinic acid from sugarcane bagasse or, in Mexico, from agave tequilana bagasse (Ávila Lara 2019). Although the production of this compound from lignocellulosic biomass is promising, there is a presence of recalcitrance, and high pretreatment costs are required. Fermentable sugars degradation using *Myceliophthora thermophila* to obtain malic and succinic acid through a consolidated bioprocessing strategy (CBP) in a single reaction system has been reported to solve these incomes (Li et al. 2020). So, a direct and selective biotechnological conversion of raw materials into products in a single unit of operation (fermentation), lower capital and energy costs lower indirect costs, resulting in safer and more sustainable products and residues, and processes (Barton et al. 2015).

Fumaric acid ($C_4H_4O_4$) is an acid like the previous one, composed of four carbons. It has an acidifying capacity 1.5 times greater than citric acid; therefore, it is commonly used in the food industry and as an ingredient for beverages (Yang ST. et al. 2011). The biological production of fumaric acid has been studied over time, but this from biorefineries is being replaced by petrochemical synthesis methods in the market (Fang, Smith, and Qui 2017). Biotechnological processes involving the use of species such as *Rhizopus* or modifications of *E. coli* using the Tricarboxylic acid cycle (TCA pathway) *S. cerevisiae* have been engineered for fumaric acid production (Fang, Smith, and Qui 2017).

Finally, within this category, we find malic acid, which is currently responsible for producing fumaric acid by catalytic maleic isomerization in aqueous solutions at low pH (Y. Liu et al. 2015).

Malic acid is extensively used in the food industry as a flavor enhancer and acidulant, providing routes to a wide range of different products and chemicals, including polymeric resins from renewable biomass (Fang, Smith, and Qui 2017). In 2008, Sauer (Sauer et al. 2008) predicted that the utilization of malic acid would be above 200,000 tons/year. In the industries, it represents a high potential as raw material for the manufacture of polymalic acid (PMA), which is a new biodegradable polymer similar to polylactic acid, but its production from malic acid depends on the economic feasibility and manufacturing process (Fang, Smith, and Qui 2017).

4.7. Biohydrocarbons: Isoprene

Compounds are classified as isoprenes (2-methyl-1,3-butadiene) if they have isopentenyl diphosphate (IPP) or dimethylallyl diphosphate (DMAPP) as a unitary molecule (Hong and Nielsen 2012). This is the largest biogenic volatile compound in the terrestrial biosphere and reaches emissions of up to 500 Tg (C) per year, which contributes significantly to climate change problems (Murrell, McGenity, and Crombie 2020). Despite environmental concerns, the isoprene market is projected to register a CAGR of over 7%, reaching a value of approximately USD 4 billion by 2025, and is expected to increase in Latin American countries such as Chile and Brazil due to construction activities (Market Research Future. 2020).

On the other hand, microorganisms have been studied for cleaner production of this compound, such as *E. coli* or the use of *Saccharomyces cerevisiae* by biosynthesis, employing a simultaneous enhancement of precursor supply and conversion through a combination of pathway compartmentalization and protein engineering (Yao et al. 2018).

Another microorganism used to obtain isoprene is *Rhodotorula glutinis*, which, through fermentation processes, has awakened great interest in coffee-producing countries for the utilization of the large volumes of residues generated by the production of isoprene (Hughes et al. 2014). The isoprene resulting from these ecological processes is a crucial molecule for obtaining other biomolecules such as sterols, hopanoids, archaeal lipids, carotenoids, chlorophyll, quinones, many signaling molecules, and hormones (Murrell, McGenity, and Crombie 2020).

4.8. Levulinic Acid (LA)

This molecule is the result of the rehydration of HMF under acidic conditions, which is endowed with a wide functionality and reactivity due to the presence of ketone and carboxylic groups (Chen et al. 2017), in the acid-catalyzed conversion of hexose sugars to levulinic acid, formic acid is co-produced as a low-value chemical used in the production of formaldehyde, rubber, plasticizers, pharmaceuticals, and textiles (Pileidis and Titirici 2016a). In recent years, different raw materials with the potential to obtain good yields of this molecule block have been explored, as shown in Table 4.

Table 4. LA production yield from different biomasses

Biomass	[%]	Acid	T (°C)	Yield [%]	Reference
Fructose	3,6-7,2	HCl	95	81	(Pileidis and Titirici 2016a)
Sugarcane bagasse	1-5	H_2SO_4	190 -220	70	(Rodríguez-Aguilera et al. 2020)
Chlorophyta microalgae	0-4	H_2SO_4	150	~30%	(Rengel et al. 2022)
Pulp residues	1-5	H_2SO_4	210-230	70-80	(Peng et al. 2010)
Wheat straw	3,5	H_2SO_4	200	68,8	(Peng et al. 2010)
Sorghum grain	8	H_2SO_4	200	45,6	(Peng et al. 2010)
Glucose	Me	SO_3H	180	41	(Peng et al. 2010)
fructose	4-Me	$C_6H_4SO_3H$	240	33,7	(Peng et al. 2010)
Bovine manure	3	HCl	180	33,8	(Jialei Su et al. 2017)

Thus, in the constant study of the products obtained, the most notable from LA is gamma-valerolactone, which is used as a green solvent, an excellent additive for fuels, and a precursor of valuable chemical products (Dutta et al. 2019). Another important application is using wastes such as red algae (*Gracilaria lemaneiformis*) to produce hydrocarbons, giving yields of approximately 45 to 55% (Dutta et al. 2019).

Ethyl levulinate is another product of interest obtained from this molecule (Lei et al. 2016). This product was obtained from biomass and was used as additional fuel to biodiesel and diesel, demonstrating properties similar to pure diesel but with better cold flow properties, in addition to complying with the standard line (B5) for blends of this type of product (Lei et al. 2016). It should be noted that essential compounds such as acrylic acid, 1,4-pentadiol, α-angelica lactone, 2-methyltetrahydrofuran (MTHF), d-aminolevulinic acid (DALA) are also obtained.

Biofine is one of the most promising processes for obtaining this compound. This process consists of acid hydrolysis of cellulose using sulfuric acid (Rodríguez-Aguilera et al. 2020), to convert it into levulinic acid and formic acid. If the raw material subjected to this process has a significant hemicellulose content, it also results in furfural. This technology allows using various raw materials, including those of low cost or some classified as waste (Susana Alcaraz Argumánez. 2013).

4.9. Sorbitol

Sorbitol is a compound used in the food, cosmetic and pharmaceutical industries (Isikgor and Becer 2015). It is produced commercially from biomass, especially that with the presence of monomeric sugars, glucose, or starch precursors, all using chemical routes (Kapanji, Haigh, and Görgens 2019), such as hydrogenation by hydrolysis of cellulose into saccharides and sugar alcohols (Zhang et al. 2015; Jofre et al. 2022).

The production of this platform molecule has aroused interest in the search for raw materials with good profitability on a commercial scale, and pine has been identified as an alternative to this problem (Rodriguez-Fernandez et al. 2022). The industrial status worldwide positions it with a TRL between 8-9 of industrial development, and the leading players are the ADM and Roquette companies, who occupy important markets for the chemical conversion of sugars, with no petrochemical alternatives (E4tech (UK) Ltd. 2017). Sorbitol has been investigated to improve the physical and mechanical properties of films made from sorghum extract, giving results comparable to petroleum-based plastics (Sari et al. 2021). Also, sorbitol was used mixed with borax to improve the heat transport in starch films from cassava (Franco-Bacca et al. 2021). On the other hand, this compound has been obtained by heterogeneous catalysis using Ru/P containing porous biochar (Mao et al. 2020).

4.10. Xylitol

Xylitol is a block molecule produced by chemical synthesis through the catalytic hydrogenation of xylose, using high temperature and pressure conditions, which leads to high energy costs, making it a more expensive product than other sugar alcohols and sugars (Queiroz et al. 2022). Accordingly, new biotechnological routes have been explored, which are mainly based on the pretreatment of the lignocellulosic biomass, followed by the concentration of the hydrolysate and detoxification; the xylose is converted into xylitol for its subsequent recovery (Felipe Hernández-Pérez et al. 2019). Some authors have studied the hydrolyzate of different lignocellulosic residues to determine the xylose content, such as sugarcane bagasse with contents of 15.73 g/L (de Arruda et al. 2011), sugarcane Straw with 18.6 g/L (Hernández-Pérez, de Arruda, and Felipe 2016), corncob with contents of 65 g/L (Kogje and Ghosalkar 2017), oil palm from bagasse 22,1 g/L (Abdul Manaf et al. 2018), the results are the starting point for the utilization and improvement of production costs of the industries dedicated to this particular sector.

Recently, multi-product installations of xylitol and glutamic acid from a 1G2G configuration (molasses and sugar cane lignocellulose feedstock) from a sugar mill have been studied. This may result in significant economic returns as a result of the implementation of mature technology for xylitol (Morakile et al. 2022), which may be related to the ability to use hemicellulose hydrolysate from raw biomass (the primary source of xylose) as a starting material, instead of isolated and purified xylose (Bozell and Petersen 2010). The use of multi-product biorefinery systems using xylitol has become a focus of study, especially for the cost-benefit analysis related to their installation (Servian-Rivas et al. 2022). (Servian-Rivas et al. 2022) analyzed a system fed by 100 tons/day of olive pruning as raw material, resulting in a positive investment balance and lower environmental impacts (0,435 kg de $CO_{2\ eq}$ global warning, 0,007 kg 1,4-DCB in human carcinogenic toxicity, 0,148 kg de Oil_{Eq}) compared to the usual production of these compounds (7,96 kg de $CO_{2\ eq}$, 0,444 kg 1,4-DCB y 2,228 kg de Oil_{Eq}).

5. Raw Materials Integration

In the context of biorefineries, integration occurs when we have a production facility in which a variety of products and by-products, ranging from fuels to chemical products with high added value, are produced from different biomasses (Sara Piedrahita Rodriguez. 2020; Zhang et al. 2018), which improves the economic feasibility of third-generation biofuel production (García Prieto 2018).

Depending on the feedstock and processes, biorefineries are classified into first, second, and third generation biorefineries (Moncada B, Aristizábal M, and Cardona A 2016). The first generation mainly used raw materials from crops such as wheat, corn, sugar beet, and sugar cane (Moncada B, Aristizábal M, and Cardona A 2016), crops that are mainly used for food and feed, as shown in Figure 6a, where the processing of sugar cane to generate bioethanol through a fermentation process can be seen.

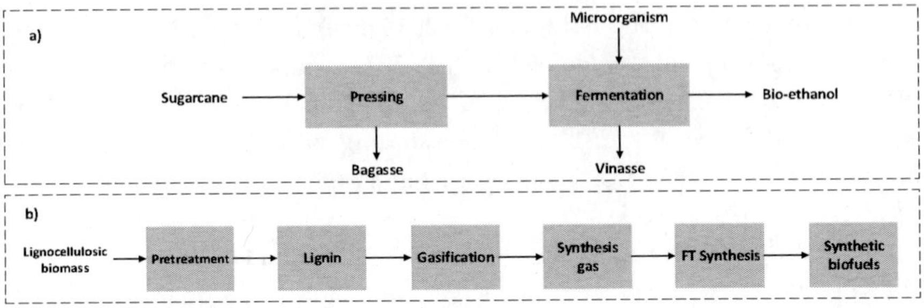

Figure 6. a) First generation biorefinery and b) Second generation biorefinery.

Second generation plants are provided with lignocellulosic material, as shown in Figure 6b, which mainly highlights the use of pretreatment of the biomass used because it tends to be recalcitrant in nature, which prevents the release of the compounds of interest. This generation of raw materials has been the focus of much research due to the conflict that exists over the use of first-generation raw materials for non-food purposes (Moncada B, Aristizábal M, and Cardona A 2016). These are rich in cellulose, which in turn can be hydrolyzed to glucose, resulting in second generation sugars, a feasible option for the valorization of lignocellulosic fractions (Bello et al. 2018; Bello et al. 2021) seeking synergies between the chemical and fuel sectors and the forestry, pulp and paper and agri-food industries (Aristizábal-Marulanda Valentina., Cardona-Alzate, and Martín 2019).

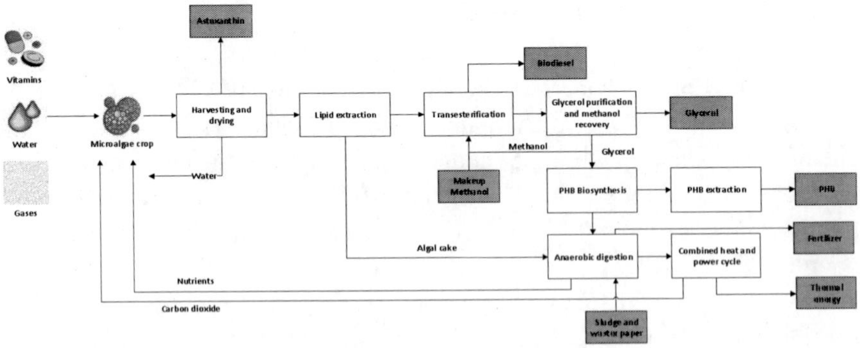

Figure 7. Integrated biorefinery from microalgae. Adapted from (García Prieto 2018).

Finally, the third generation is based on algae or cyanobacteria containing a high oil mass fraction cultivated in ponds (Suganya et al. 2016); the use of aquatic biomass as a feedstock for refining is the challenge (Nawaj Alam,

Singh, and Guldhe 2021). An example of this type of third generation biorefineries to produce biodiesel and high-value-added products is shown in Figure 7, where, apart from bioethanol, the production of biopolymers (PHB) and nutraceuticals is included (astaxantina) (García Prieto 2018).

6. Energy Integration

Using non-renewable energy sources to meet the world's energy needs has caused various problems, such as global warming, water pollution, and smog production. In this sense, lignocellulosic biomass has been postulated as a renewable energy source capable of producing energy carriers that can cover this energy demand, given that it does not represent a danger to the food security of humankind. Studies conducted by (Solarte-Toro, Chacón-Pérez, and Cardona-Alzate 2018) indicated that biogas and syngas are two energy vectors that have been proposed to generate heat and electricity through their use in cogeneration systems, comparisons have been made between these energy vectors considering their main characteristics and a technical-economic and energetic evaluation of heat and electricity generation using these vectors as energy sources has been performed.

If lignocellulosic biomass is used as feedstock, biogas is more commonly used for cogeneration purposes than syngas. However, syngas from biomass gasification has great potential to be used as a chemical platform to produce value-added products. However, the selection of one technology or the other is also subject to the intrinsic characteristics of the feedstock, always requiring precise analysis of these two options at the technical level. In addition, the investment costs for generating heat and power from lignocellulosic materials using anaerobic digestion technology are higher than those using gasification technology. In conclusion, it was found that purified biogas (i.e., methane contents above 80%) has a higher potential to produce heat and power than syngas. Still, implementing both energy carriers in the energy market is essential to meet the world's growing demand for energy, as has been realized in countries such as Thailand, China, and Indonesia.

In evaluations carried out to produce bioethanol, biogas, and electricity for the use of banana pseudo-stalk and rice husks produced as agricultural waste in Colombia through a biorefinery scheme, the chemical characterizations of these wastes were used to obtain the input data for the conceptual design of the fermentation, anaerobic digestion and combustion processes (Daza Serna et al. 2016). They also developed a technical-economic evaluation to compare the potential of each process. The rice husk biorefinery showed higher biogas and electricity yields than the banana pseudostem case. In general terms, it was observed that the production costs of bioethanol (0.48 USD/kg vs. 0.82 USD/kg), biogas (0.27 USD/m^3 vs. 0.56 USD/m^3), and electricity (0.02 USD/kW vs. 0.03 USD/kW) remained low in a biorefinery approach compared to the individual production of each energy vector for the case of rice husk. In the case of plantain, all the results were negative for both the biorefinery and the banana cases. This work concluded that it is possible to take advantage of some agricultural residues generated in Colombia under the biorefinery scheme to produce bioethanol, biogas, and electricity as a source of energy to supply a fraction of the country's energy demand.

However, potential applications related to biomass conversion into energy carriers have been promoted in Colombia, considering biogas as a potential biofuel derived from agricultural waste. Parra et al. (Parra-Ramírez, Solarte-Toro, and Cardona-Alzate 2020), evaluated the simulation of the biogas production process from banana pseudostem. Three scenarios were proposed and simulated using Aspen Plus software. The first scenario only includes anaerobic degradation of banana pseudostem without pretreatment and biogas purification steps. The second scenario involves acid pretreatment of the feedstock and biogas production using the mixture between a xylose-rich liquor and the acid-pretreated solid as substrate. In contrast, the third scenario includes an acid pretreatment stage and a biogas purification technology. The results show that banana pseudostem is a promising feedstock for biogas production. The application of dilute acid pretreatment to the feedstock increases methane yields. In addition, implementing the pretreatment and methane concentration stages generates higher production costs. However, the third scenario showed better performance since it has lower carbon dioxide content. In conclusion, biogas produced from renewable sources has great potential to be used as a biofuel and energy source for electricity generation in rural areas of Colombia. On the other hand, recent research on the use of residues from the cassava value chain with the biorefinery approach has shown that the use of cassava residues in the value chain can be used as a source of energy (Zhang et al. 2016). As an important food crop, it has been widely used as a feedstock for various agricultural industries to produce starch, bioethanol, and other bio-based products/chemicals (Padi and Chimphango 2020). These cassava-based industries also generate large amounts of waste, rich in organic matter and suspended solids and pose significant environmental problems. Thailand has 92 processing plants with a total production capacity of 1910 and

4350 tons/day of native and modified starch. The operation is approximately 24h/day and 9 months (Chavalparit and Ongwandee 2009).

Their complex biochemical composition with high organic content gives them great potential for bioconversion into value-added products through biorefinery, which provides economic and environmental sustainability to cassava industries. The main products studied have been mainly biofuels (ethanol and butanol), biogas, biosurfactants, organic acids, and other valuable biochemicals. Cassava has been identified as a potential biological resource for integrated biorefineries to co-produce bioproducts [succinic acid (SA), glucose syrup (GS), bioethanol] and combined heat and power (CHP) (Padi and Chimphango 2020). Technical-economic evaluations of five biorefinery scenarios for cassava starch plant integration (200 tons/day), co-processing of 377.83 ton/h of wastewater (CWW), 7.29 ton/h of bagasse (CB), and 450.89 ton/h of stems (CS), were conducted using Aspen Plus® to determine their potential commercial viability. Scenarios (I) and (II) coprocess CB + CWW for biogas and bioethanol production, respectively, and CS for CHP. Scenario (III)-(V) coprocess CB + CWW + 10% CS for bioethanol (III), coproducing GS (IV) or SA (V) and 90% CS for CHP. All scenarios meet CHP's demands for biorefinery and starch processing. However, only the products in Scenario (V) had minimum selling prices equal to market prices. Therefore, integrating HS production (6.9 ton/h) in a biorefinery that co-produces bioethanol and cogeneration is a potentially viable cassava waste biorefinery with economic and environmental benefits (Padi and Chimphango 2020).

(Patrizi et al. 2020) evaluated the level of sustainability of two biofuel productions based on starch and lignocellulosic feedstocks (i.e., cassava peel and corn stover, respectively). A first understanding of the sustainability of the case studies was obtained, and the results showed that the cassava peel-based biorefinery was more sustainable from both the user's perspective and the waste generator. LCA results showed that the impact categories Global Warming Potential (GWP) and Acidification Potential (AP) had lower values for cassava compared to corn stover and a higher percentage of renewability (94%). These results suggest that biorefineries are an option for the global bioeconomy strategy, as they allow the optimization of agricultural and food waste and its environmental performance in producing a renewable substitute for fossil fuels and other non-renewable materials (Patrizi et al. 2020).

7. Non-Conventional Biobased Materials

In the search to generate sustainable alternatives, techniques have been developed to obtain composites from renewable raw materials (Tolosa et al. 2014), attracting attention thanks to their diverse thermal, magnetic, electric, chemical, and morphological properties and the wide range of applications they have and their ability to meet the expectations of use compared to conventional materials (Cervantes-Álvarez et al. 2014). Composite materials emerged in the middle of the 20th century as a promising class of engineering materials providing new prospects for modern technology (Vasiliev and Morozov 2018).

Among these raw materials, those known to be lignocellulosic materials stand out (Calvo et al. 2012). This is because it is an important resource for biorefining renewable energies into fuels, chemical products, and bio-based materials (Gandla, Martín, and Jönsson 2018). An example of this is the effectiveness of using raw materials such as corn and sugarcane for ethanol production, a fast-growing industry in the world, led by the United States and Brazil (Chum et al. 2014). The use of lignocellulosic materials obtained from other sources that are not intended for human consumption has become a focus of study due to the need for alternative solutions that do not jeopardize food safety and that also meet the needs required by the markets.

Non-conventional biobased materials are part of the solutions, which are intended to be sustainable for the problems caused by industries related to environmental and social issues. Accordingly, studies have been developed based on the search for renewable raw materials to mitigate many of the effects caused by using petrochemical compounds for communities, which are related to the environment and human health. An example of this situation is bisphenol-A epoxy, this bisphenol-based product is mainly used to manufacture polycarbonate bottles, the inner lining of cans, dental sealants, and even cell phone casings (Verge et al. 2014). However, medical studies have determined that it is considered an endocrine disruptor with the capacity to alter body functions and systems (Juan-García, Gallego, and Font 2015), putting at risk the health of the people who are in contact with it. Due to the warnings of its use against health, alternative materials have been sought for its production, which are safe and do not generate reactions when in contact. One of these alternatives is a derivative of cashew nutshell liquid, whose main component is cardanol, an unsaturated metaalkylphenol, which, being a by-product of cashew shell processing, does not put at risk the production of food resources and, according to the first experimental advances, has resulted in an adhesive with potential for the replacement of bisphenol-A epoxy with acceptable capabilities as a coating (Verge et al. 2014).

Thus, using renewable materials has been the starting point for obtaining compounds that meet market needs without jeopardizing the aspects as mentioned above. Within the topic of polymers, approaches have been developed towards their elaboration based on 2,5-furanodicarboxylic acid (2,5-FDCA), which constitutes a new class of aliphatic polyesters that can be prepared from monomers derived directly from renewable resources such as furfural and hydroxymethylfurfural (HMF) (Papageorgiou et al. 2016). Combinations of other molecules such as succinic acid, 2,5-FDCA, 5,50-(propane-2,2-diyl)-bis(furan-2-carboxylate), and 1,3-propanediol, have also yielded results obtaining amorphous and semi-crystalline copolyesters, which depending on the proportion of monomers, resulting in materials with high thermal stability (Hbaieb et al. 2015). The potential of using erythritol dicarbonate and bio-based diamines such as butaneido dicarbonate to produce polyhydroxyurethane compounds (Schmidt et al. 2017), making it a competitive process with conventional reactive processing of polyurethane thermoplastics using isocyanates. Bioplastics are biodegradable materials from renewable sources.

In the construction sector, the reduction of energy consumption by equipment such as mortars has been sought (Giosuè et al. 2016); in water-sand mortars with cement, the substitution of non-conventional aggregates was used to replace water and sand with hydraulic lime and spruce chips and the ashes of the same biomass retained in the equipment. This increased the mechanical performance of the mortars, and the biomass-based ashes could generate better effects in terms of water vapor permeability and make the mortars lighter in their performance. Other materials have been identified as possible alternatives in this industry, such as the case of date palm fibers, which, together with cement and sand in mortars, turn them into bio-based equipment with excellent moisture buffering capacity (Chennouf et al. 2018). The advantage of using bio-based building materials is retaining carbon dioxide, which could preserve buildings for a longer time (Pacheco-Torgal 2020). Synthetic composites have displaced the use of natural materials such as wool (Rajabinejad, Bucişcanu, and Maier 2019), the alternatives for the elaboration of compounds from the residues left over from the breeding farms of animals with this coat. They identify mainly the potential of the extraction of the keratin contained in them by chemical solubilization or enzymatic bioconversion, for the production of protein-based biomaterials for the generation of different medicines (Rajabinejad, Bucişcanu, and Maier 2019).

Biochar is a byproduct of energy production through pyrolysis or gasification. Biochar is considered a nanoporous material (Dzyazko et al. 2019), used mainly as adsorbents (Ambaye et al. 2021) and electrodes for energy storage devices (Senthil and Lee 2021).

On the other hand, it is possible to obtain biogenic silica as by products to the extraction of nanocellulose from pineapple peels (Corrales-Ureña et al. 2018) or rice husk (Steven, Restiawaty, and Bindar 2021), with the potential applications as reinforcement in adhesives, polymer composites, in the biomedical field, and even as a source of silica for fertilizers (Corrales-Ureña et al. 2018).

8. International Market and Future Challenges

Biorefinery concepts rely more on their ability to recover chemicals with high value and market potential (Awasthi et al. 2022) to meet the needs of the 21st century, such as the imminent depletion of fossil fuels and climate change issues (Shahid et al. 2021). Because of this, the use of biomass as a fuel and energy source has been chosen, although this is where the first challenge begins. There is a limitation in supplies and maximum conversion yields (Ubando, Felix, and Chen 2020), although in recent years, novel technologies have emerged to improve processes and make them sustainable (Yong et al. 2022). Another problem is the disposal and transportation of some raw materials, especially in Latin American countries, where conflicts and inadequate management make logistics costly, unsafe, and unreliable (Perez. 2017), increasing final costs and often limiting production processes.

Accordingly, not only the layout and the way of processing would be challenges, but also the understanding of the concepts related to the term biorefineries. Globally, this has many interpretations, just like the circular economy, and this has caused it to remain at a dead end due to a permanent conceptual dispute, not only in research but also in practice, as the development of cumulative knowledge is independent (Kirchherr, Reike, and Hekkert 2017). Although bioenergy is an important vector obtained from biorefinery in the energetic transition, value-added products such as biopolymers, antioxidants, and essential oils are recognized as the driven potential for international markets such as European Union (Orejuela Escobar 2018; Del Pozo et al. 2020; Potrč et al. 2020). These value-added products are required by the cosmetic and pharmaceutical industries (Gómez-Cruz et al. 2021).

Another opportunity to be fully exploited within these products obtained from biorefinery schemes is the production of ferulic acid (Sharma et al. 2020) from residues such as corn residues, which, subject to direct extraction processes, result in obtaining vanillin and guaiacol, important products for the food industry (Peña and Lopez 2020). From the non-food sector, from this monomer, it would be possible to replace some petroleum-based methacrylate for molding production, making it a promising compound for producing more environmentally friendly materials (Peña and Lopez 2020). These composites are mentioned because they are part of the future challenges facing these emerging industries: research and development of new composites from biomass is key to the sustainable development of markets seeking to replace petroleum-based composites, although a limitation to this could be reflected in the lack of funding resources for innovation, specifically in Latin America (Rodríguez, Sotomayor, and Rodrigues 2019). Public funds are scarce and generally limited to financing the initial phases of research and development; the venture capital market is incipient in most countries; and there is no widespread culture of investment in innovation by the private sector (Rodríguez, Sotomayor, and Rodrigues 2019).

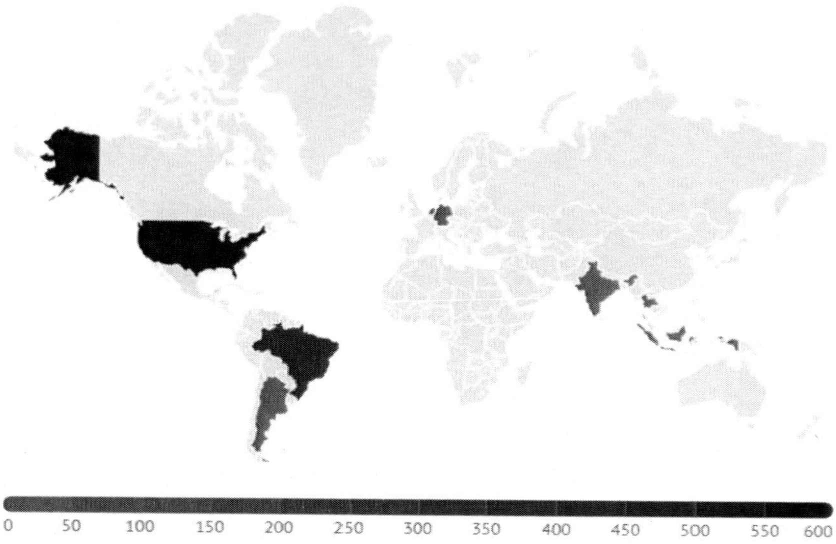

Figure 8. Distribution of the leading producers of biofuels for the year 2020. Production in thousands of barrels of oil equivalent per day.

On the other hand, the market share of renewables in global electricity generation reached almost 29% in 2020, where there was a record percentage increase, which could be directly related to the slowdown in economic activity due to Covid-19. Despite the increase, there is still a long way to go to meet the proposed scenario of net zero emissions by 2050 of more than 60% of generation by 2030 (Bioenergy. 2022). Demand is also expected to grow by 28% during 2021 - 2026, which is equivalent to 41 billion liters, with the United States and Brazil as the leading suppliers of this demand, as shown in Figure 8 (IEA Bioenergy. 2021).

Implementing biorefineries fed with third generation feedstock is seen as the best alternative to meet global energy goals. The future of microalgae use is coming into the limelight due to market prices and the fact that many of the current biomass conversion techniques are not cost-effective or viable in the long term (Thanigaivel, Vickram, et al. 2022). However, many bottlenecks and challenges are associated with production processes and downstream strategies, which are difficult to manage economically (Prabha et al. 2022). These approaches, accompanied by public policies and financing funds, will allow the proper development of future research related to biorefineries, to obtain multiple products with high added value and, above all, with economic benefits that encompass all stakeholders without neglecting the gains of the planet earth and its conservation (Solarte-Toro and Cardona Alzate 2021a; Giraldo et al. 2020).

Conclusion

The future of biorefineries must be aligned to the bioeconomy framework and the four pillars of sustainability (Solarte-Toro and Cardona Alzate 2021b). Therefore, it is necessary to conduct systematic, transversal research to explore the new platform molecules and strategies for raw or energy integration to improve the integral use of biomass and increase revenues.

By joining efforts with the developments provided by nanotechnology, it is possible to increase the added value of biorefineries. Thanigaivel et al. reports the connection between nanotechnology and biorefinery (Thanigaivel, Priya, et al. 2022). Lignocellulosic biomass is a nanometer-sized at the atomic level, and therefore it acts as a fundamental building block that results in achievement of their extraordinary properties in various commercial applications (Thanigaivel, Priya, et al. 2022). Renewable materials and lignocellulosic biomass nanotechnology are critical to improve functioning while minimizing environmental impact. Nanostructured materials have been used to obtain biofuels, such as biodiesel. Lipase immobilized on magnetite-covered silica shows an increase in conversion to biodiesel compared with free enzyme (Kalantari, Kazemeini, and Arpanaei 2013). However, additional research efforts in Latin America countries have been done to close the poverty gap and the economic impact of the Sars-Cov 2 pandemic.

The biorefineries implementation and viability depend strongly on technological maturity, biomass availability (high volumes), location and access of residue sources (logistic issues), topography of rural zones, agricultural and cultural practices to change traditional uses of biomass, and adaptive capacity to an energy transition on the part of those countries highly dependent on fossil resources.

Acknowledgments

The authors express their gratitude to the project research called "Aprovechamiento y valorización sostenible de residuos sólidos orgánicos y su posible aplicación en borrefinerías y tecnologías de residuos a energía en el departamento de Sucre" with BPIN code 2020000100189, funded by the Sistema General de Regalías - SGR.

References

Abdul Manaf, Shareena Fairuz, Jamaliah Md Jahim, Shuhaida Harun, and Abdullah Amru Indera Luthfi. 2018. "Fractionation of oil palm fronds (OPF) hemicellulose using dilute nitric acid for fermentative production of xylitol." *Industrial Crops and Products* 115:6-15. doi: 10.1016/j.indcrop.2018.01.067.

Abdulkhani, Ali, Meraj Siahrang, Zahra Echresh Zadeh, Sahab Hedjazi, Sanaz Torkameh, and Mehdi Faezipour. 2021. "Direct catalytic conversion of bagasse fibers to furan building blocks in organic and ionic solvents." *Biomass Conversion and Biorefinery*. doi: 10.1007/s13399-021-01592-w.

Ahmad, Farah B., M.A. Kalam, Zhanying Zhang, and H.H. Masjuki. 2022. "Sustainable production of furan-based oxygenated fuel additives from pentose-rich biomass residues." *Energy Conversion and Management: X* 14:100222. doi: 10.1016/j.ecmx.2022.100222.

Alexis Michael Bazzanella, Florian Ausfelder, and DECHEMA Gesellschaft für Chemische Technik und Biotechnologie e.V. 2017. [DECHEMA Society for Chemical Engineering and Biotechnology e.V. 2017.]. *Low carbon energy and feedstock for the European chemical industry*. Edited by The European Chemical Industry Council.

Ambaye, T. G., M. Vaccari, E. D. van Hullebusch, A. Amrane, and S. Rtimi. 2021. "Mechanisms and adsorption capacities of biochar for the removal of organic and inorganic pollutants from industrial wastewater." *International Journal of Environmental Science and Technology* 18 (10):3273-3294. doi: 10.1007/s13762-020-03060-w.

Andler, R., C. Valdés, V. Urtuvia, C. Andreeßen, and A. Díaz-Barrera. 2021. "Fruit residues as a sustainable feedstock for the production of bacterial polyhydroxyalkanoates." *Journal of Cleaner Production* 307:127236. doi: https://doi.org/10.1016/j.jclepro.2021.127236.

Aristizábal-Marulanda Valentina., Carlos Ariel. Cardona- Alzate, and Mariano. Martín. 2019. "An integral methodological approach for biorefineries design: Study case of Colombian coffee cut-stems." *Computers & Chemical Engineering* 126:35-53. doi: https://doi.org/10.1016/j.compchemeng.2019.03.038.

Aristizábal-Marulanda, Valentina, Juan Camilo Solarte-Toro, and Carlos Ariel Cardona Alzate. 2021. "Study of biorefineries based on experimental data: production of bioethanol, biogas, syngas, and electricity using coffee-cut stems as raw material." *Environmental Science and Pollution Research* 28 (19):24590-24604. doi: 10.1007/s11356-020-09804-y.

Aristizábal-Marulanda, Valentina., Christian D. Botero-Gutierrez, and Carlos A. Cardona-Alzate. 2019. "Chapter 4 - Thermochemical, Biological, Biochemical, and Hybrid Conversion Methods of Bio-derived Molecules into Renewable Fuels." In *Advanced Bioprocessing for Alternative Fuels, Biobased Chemicals, and Bioproducts*, edited by Majid Hosseini, 59-81. Woodhead Publishing.

Ávila Lara, Abimael Iván. 2019. *"Evaluación de bagazo de Agave tequilana Weber variedad azul y de Opuntia ficus indica en pretratamientos químicos y enzimáticos para su aplicación en la producción de ácido succínico."* [*Evaluation of bagasse from Agave tequilana Weber blue variety and from Opuntia ficus indica in chemical and enzymatic pretreatments for its application in the production of succinic acid*]. Maestría Maestría, Maestría en Ciencias con Orientación en Procesos Sustentables, Universidad Autónoma de Nuevo León. [Master's Degree, Master's Degree in Sciences with Orientation in Sustainable Processes, Autonomous University of Nuevo León].

Awasthi, Mukesh Kumar, Raveendran Sindhu, Ranjna Sirohi, Vinod Kumar, Vivek Ahluwalia, Parameswaran Binod, Ankita Juneja, Deepak Kumar, Binghua Yan, Surendra Sarsaiya, Zengqiang Zhang, Ashok Pandey, and Mohammad J. Taherzadeh. 2022. "Agricultural waste biorefinery development towards circular bioeconomy." *Renewable and Sustainable Energy Reviews* 158:112122. doi: 10.1016/j.rser.2022.112122.

Azlan, Nadiah Syafiqah Mohd, Chiew Lin Yap, Suyin Gan, and Mohd Basyaruddin Abdul Rahman. 2022. "Recent advances in the conversion of lignocellulosic biomass and its degraded products to levulinic acid: A synergy of Brønsted-Lowry acid and Lewis acid." *Industrial Crops and Products* 181:114778. doi: https://doi.org/10.1016/j.indcrop.2022.114778.

Barton, Nelson R., Anthony P. Burgard, Mark J. Burk, Jason S. Crater, Robin E. Osterhout, Priti Pharkya, Brian A. Steer, Jun Sun, John D. Trawick, Stephen J. Van Dien, Tae Hoon Yang, and Harry Yim. 2015. "An integrated biotechnology platform for developing sustainable chemical processes." *Journal of Industrial Microbiology and Biotechnology* 42:349-360. doi: 10.1007/s10295-014-1541-1.

Beatriz Briano Col. 2021. *Valorizacion de residuos en la industria carnica bovina*. [*Waste recovery in the bovine meat industry*]. https://www.cyted.org/sites/default/files/c19-valorizacion_de_residuos_industria_carnica-beatriz_briano-18-oct.pdf. CYTED.

Becker, Judith, Anna Lange, Jonathan Fabarius, and Christoph Wittmann. 2015. "Top value platform chemicals: Bio-based production of organic acids." *Current Opinion in Biotechnology* 36:168-175. doi: 10.1016/J.COPBIO.2015.08.022.

Bello, Sara, Carmen Ríos, Gumersindo Feijoo, and María Teresa Moreira. 2018. "Comparative evaluation of lignocellulosic biorefinery scenarios under a life-cycle assessment approach." *Biofuels, Bioproducts and Biorefining* 12 (6):1047-1064. doi: https://doi.org/10.1002/bbb.1921.

Bello, Sara, Noelia Pérez, Jan Kiebist, Katrin Scheibner, María Isabel Sánchez Ruiz, Ana Serrano, Ángel T. Martínez, Gumersindo Feijoo, and Maria Teresa Moreira. 2021. "Early-stage sustainability assessment of enzyme production in the framework of lignocellulosic biorefinery." *Journal of Cleaner Production* 285:125461. doi: https://doi.org/10.1016/j.jclepro.2020.125461.

Benites-Lazaro, L. L., L. L. Giatti, W. C. Sousa Junior, and A. Giarolla. 2020. "Land-water-food nexus of biofuels: Discourse and policy debates in Brazil." *Environmental Development* 33:100491. doi: https://doi.org/10.1016/j.envdev.2019.100491.

Bhatia, Shashi Kant, Sujit Sadashiv Jagtap, Ashwini Ashok Bedekar, Ravi Kant Bhatia, Anil Kumar Patel, Deepak Pant, J. Rajesh Banu, Christopher V. Rao, Yun Gon Kim, and Yung Hun Yang. 2020. "Recent developments in pretreatment technologies on lignocellulosic biomass: Effect of key parameters, technological improvements, and challenges." *Bioresource Technology* 300. doi: 10.1016/J.BIORTECH.2019.122724.

Bio Costa Rica. 2022. "*Estrategia Nacional de Bioeconomía Costa Rica 2030-2050*. [*National Bioeconomy Strategy Costa Rica 2030-2050*]. https://www.chmcostarica.go.cr/recursos/documentos-y-publicaciones/estrategia-nacional-de-bioeconomia-costa-rica-2020-2030. accesed on july 2022." https://www.chmcostarica.go.cr/recursos/documentos-y-publicaciones/estrategia-nacional-de-bioeconomia-costa-rica-2020-2030.

Bioenergy., IEA. 2022. "Renewables." *Renewables, including solar, wind, hydro, biofuels and others, are at the centre of the transition to a less carbon-intensive and more sustainable energy system.*

BIOFIN. 2021. *"Estrategia de Bioeconomía de BIOFIN México*. [*BIOFIN Mexico Bioeconomy Strategy*]. https://www.biofin.org/knowledge-product/estrategia-de-bioeconomia-de-biofin-mexico. accessed july 2022."

Birgit Kamm, and Michael Kamm. 2007. "International biorefinery systems." *Pure and Applied Chemsitry* 79 (11):1983-1997. doi: doi:10.1351/pac200779111983.

Borges, E. R., and N. Pereira Jr. 2011. "Succinic acid production from sugarcane bagasse hemicellulose hydrolysate by Actinobacillus succinogenes." *Journal of Industrial Microbiology and Biotechnology* 38:1001-1011. doi: 10.1007/s10295-010-0874-7.

Bozell, Joseph J., and Gene R. Petersen. 2010. "Technology development for the production of biobased products from biorefinery carbohydrates—the US Department of Energy's "top 10" revisited." *Green Chemistry* 12:539-55. doi: 10.1039/B922014C.

Bragatto, J., F. Segato, and F.M Squina. 2013. "Production of xylooligosaccharides (XOS) from delignified sugarcane bagasse by peroxide-HAc process using recombinant xylanase from Bacillus subtilis." *Industrial Crops and Products* 51:123-129. doi: 10.1016/j.indcrop.2013.08.062.

Brownlee, Harold J., and Carl S. Miner. 1948. "Industrial Development of Furfural." *Industrial & Engineering Chemistry* 40 (2):201-204. doi: 10.1021/ie50458a005.

Bundhoo, M., R. Mohee, and M. Hassan. 2015. "Effects of pre-treatment technologies on dark fermentative biohydrogen production: A review." *Journal of Environmental Management* 157:20-48. doi: https://doi.org/10.1016/j.jenvman.2015.04.006.

Bustamante, Daniel, Marta Tortajada, Daniel Ramón, and Antonia Rojas. 2019. "Production of D-Lactic Acid by the Fermentation of Orange Peel Waste Hydrolysate by Lactic Acid Bacteria." *Fermentation,* 6(1), 1.doi: 10.3390/fermentation6010001.

Caderby, E., S. Baumberger, W. Hoareau, C. Fargues, M. Decloux, and M. Maillard. 2013. "Sugar Cane Stillage: A Potential Source of Natural Antioxidants." *Journal of Agricultural and Food Chemistry* 61. doi: 10.1021/jf4039474.

Calvo, S., N. P. Arias, O. Giraldo, A. Rosales-Rivera, and O. Moscoso. 2012. "Thermal and magnetic behavior of Angustifolia Kunth bamboo fibers covered with Fe_3O_4 particles." *Physica B: Condensed Matter* 407 (16):3267-3270. doi: https://doi.org/10.1016/j.physb.2011.12.083.

Carbonell, Sergio A. M., Luis Augusto Barbosa Cortez, Luis F. C. Madi, Lilian C. Anefalos, Ricardo Baldassin Junior, and Rodrigo L. V. Leal. 2021. "Bioeconomy in Brazil: Opportunities and guidelines for research and public policy for regional development." *Biofuels, Bioproducts and Biorefining* 15 (6):1675-1695. doi: https://doi.org/10.1002/bbb.2263.

Castillo Martinez, Fabio Andres, Eduardo Marcos Balciunas, José Manuel Salgado, José Manuel Domínguez González, Attilio Converti, and Ricardo Pinheiro de Souza Oliveira. 2013. "Lactic acid properties, applications and production: A review." *Trends in Food Science and Technology* 30:70-83. doi: 10.1016/J.TIFS.2012.11.007.

Cervantes-Álvarez, F., J. J. Reyes-Salgado, V. Dossetti, and J. L. Carrillo. 2014. "Thermal properties of composite materials with a complex fractal structure." *Journal of Physics D: Applied Physics* 47 (23):235303. doi: 10.1088/0022-3727/47/23/235303.

Chavalparit, Orathai, and Maneerat Ongwandee. 2009. "Clean technology for the tapioca starch industry in Thailand." *Journal of Cleaner Production* 17 (2):105-110. doi: https://doi.org/10.1016/j.jclepro.2008.03.001.

Chen, Chunlin, Lingchen Wang, Bin Zhu, Zhenqiang Zhou, Soliman I. El-Hout, Jie Yang, and Jian Zhang. 2021. "2,5-Furandicarboxylic acid production via catalytic oxidation of 5-hydroxymethylfurfural: Catalysts, processes and reaction mechanism." *Journal of Energy Chemistry* 54:528-554. doi: 10.1016/j.jechem.2020.05.068.

Chen, Season S., Thawatchai Maneerung, Daniel C.W. Tsang, Yong Sik Ok, and Chi-Hwa Wang. 2017. "Valorization of biomass to hydroxymethylfurfural, levulinic acid, and fatty acid methyl ester by heterogeneous catalysts." *Chemical Engineering Journal* 328:246-273. doi: 10.1016/j.cej.2017.07.020.

Chennouf, Nawal, Boudjemma Agoudjil, Abderrahim Boudenne, Karim Benzarti, and Fathi Bouras. 2018. "Hygrothermal characterization of a new bio-based construction material: Concrete reinforced with date palm fibers." *Construction and Building Materials* 192:348-356. doi: 10.1016/j.conbuildmat.2018.10.089.

Cherubini, Francesco, Gerfried Jungmeier, Maria Wellisch, Thomas Willke, Ioannis Skiadas, René Van Ree, and Ed de Jong. 2009. "Toward a common classification approach for biorefinery systems." *Biofuels, Bioproducts and Biorefining,* 3(5), 534–546. doi: https://doi.org/10.1002/bbb.172.

Choi, Sol, Chan Woo Song, Jae Ho Shin, and Sang Yup Lee. 2015. "Biorefineries for the production of top building block chemicals and their derivatives." *Metabolic Engineering* 28:223-239. doi: 10.1016/J.YMBEN.2014.12.007.

Chum, Helena L., Ethan Warner, Joaquim E.A. Seabra, and Isaias C. Macedo. 2014. "A comparison of commercial ethanol production systems from Brazilian sugarcane and US corn." *Biofuels, Bioproducts and Biorefining* 8:205-223. doi: 10.1002/bbb.1448.

Chung, Millicent Rosette Wan Yi, Inn Shi Tan, Henry Chee Yew Foo, Man Kee Lam, and Steven Lim. 2021. "Potential of macroalgae-based biorefinery for lactic acid production from exergy aspect." *Biomass Conversion and Biorefinery*. doi: 10.1007/S13399-021-01375-3.

Commission to the Gothenburg European Council. 2005. "*Strategy for sustainable development.*" accessed june 20. https://eur-lex.europa.eu/legal-content/EN/TXT/HTML/?uri=URISERV:l28117&from=DE.

CONPES 2834. 1996. CONPES 2834. *Política de Bosques.* [*Forest Policy*]. https://colaboracion.dnp.gov.co/CDT/Conpes/Econ%C3%B3micos/2834.pdf.

CONPES 3510. 2008. CONPES 3510. *Lineamientos de política para promover la producción sostenible de biocombustibles en Colombia.* [*Policy guidelines to promote the sustainable production of biofuels in Colombia*]. https://colaboracion.dnp.gov.co/CDT/Conpes/Econ%C3%B3micos/3510.pdf.

CONPES 3527. 2008. CONPES 3527. *Política Nacional de Competitividad y Productividad Política Nacional de Competitividad y Productividad.* [*National Competitiveness and Productivity Policy National Competitiveness and Productivity Policy*]. https://colaboracion.dnp.gov.co/CDT/Conpes/Econ%C3%B3micos/3527.pdf.

CONPES 3533. 2008. CONPES 3533. *Bases de un Plan de Acción para la Adecuación del Sistema de Propiedad Intelectual a la Competitividad y Productividad Nacional 2008-2010.* [*Bases of an Action Plan for the Adaptation of the Intellectual Property System to National Competitiveness and Productivity 2008-2010*]. https://colaboracion.dnp.gov.co/CDT/Conpes/Econ%C3%B3micos/3533.pdf.

CONPES 3582. 2009. CONPES 3582. *Política Nacional de Ciencia, Tecnología e Innovación.* [*National Science, Technology and Innovation Policy*]. https://colaboracion.dnp.gov.co/CDT/Conpes/Econ%C3%B3micos/3582.pdf.

CONPES 3678. 2010. CONPES 3678 *Política de Transformación Productiva: Un Modelo de Desarrollo Sectorial para Colombia.* [*Productive Transformation Policy: A Sectoral Development Model for Colombia*] https://colaboracion.dnp.gov.co/CDT/Conpes/Econ%C3%B3micos/3678.pdf.

CONPES 3697. 2011. CONPES 3697. *Política para el Desarrollo Comercial de la Biotecnología a Partir del Uso Sostenible de la Biodiversidad.* [*Policy for the Commercial Development of Biotechnology from the Sustainable Use of Biodiversity*]. https://colaboracion.dnp.gov.co/CDT/Conpes/Econ%C3%B3micos/3697.pdf.

CONPES 3700. 2011. CONPES 3700. *Estrategia institucional para la articulación de políticas y acciones en materia de cambio climático en Colombia. [Institutional strategy for the articulation of policies and actions on climate change in Colombia].* https://colaboracion.dnp.gov.co/CDT/Conpes/Econ%C3%B3micos/3700.pdf.

CONPES 3834. 2015. CONPES 3834. *Lineamientos de política para estimular la inversión privada en ciencia, tecnología e innovación a través de deducciones tributarias.* [*Policy guidelines to stimulate private investment in science, technology and innovation through tax deductions*]. https://colaboracion.dnp.gov.co/CDT/Conpes/Econ%C3%B3micos/3834.pdf.

CONPES 3866. 2016. CONPES 3866. *Política Nacional de Desarrollo Productivo.* [*National Productive Development Policy*]. https://colaboracion.dnp.gov.co/CDT/Conpes/Econ%C3%B3micos/3866.pdf.

CONPES 3874. 2016. CONPES 3874. *Política Nacional para la Gestión Integral de Residuos Sólidos.* [*National Policy for the Integral Management of Solid Waste*]. https://colaboracion.dnp.gov.co/CDT/Conpes/Econ%C3%B3micos/3874.pdf.

CONPES 3892. 2017. CONPES 3892. *Actualización del Documento CONPES 3834 Lineamientos de política para estimular la inversión privada en ciencia, tecnología e innovación a través de deducciones tributarias.* [*Policy guidelines to stimulate private investment in science, technology and innovation through tax deductions*]. https://colaboracion.dnp.gov.co/CDT/Conpes/Econ%C3%B3micos/3892.pdf.

CONPES 3918. 2018. CONPES 3918. *Estrategia para la implementación de los Objetivos de Desarrollo Sostenible (ODS) en Colombia.* [*Strategy for the implementation of the Sustainable Development Goals (SDG) in Colombia*]. https://colaboracion.dnp.gov.co/CDT/Conpes/Econ%C3%B3micos/3918.pdf.

CONPES 3934. 2018. CONPES 3934. *Política de Crecimiento Verde.* [*Green Growth Policy*]. https://colaboracion.dnp.gov.co/CDT/Conpes/Econ%C3%B3micos/3934.pdf.

CONPES 3975. 2019. CONPES 3975. *Política Nacional para la Transformación Digital e Inteligencia Artificial.* [*National Policy for Digital Transformation and Artificial Intelligence*]. https://colaboracion.dnp.gov.co/CDT/Conpes/Econ%C3%B3micos/3975.pdf.

CONPES 3990. 2020. CONPES 3990. *Colombia potencia bioceánica sostenible 2030.* [*Colombia sustainable bi-oceanic power 2030*]. https://colaboracion.dnp.gov.co/CDT/Conpes/Econ%C3%B3micos/3990.pdf.

CONPES 4004. 2020. CONPES 4004. *Economía circular en la gestión de los servicios de agua potable y manejo de aguas residuales.* [*Circular economy in the management of *drinking water services and wastewater management*]. https://colaboracion.dnp.gov.co/CDT/Conpes/Econ%C3%B3micos/4004.pdf.

CONPES 4021. 2020. CONPES 4021. *Política Nacional para el Control de la Deforestación y la Gestión Sostenible de los Bosques.* [*National Policy for the Control of Deforestation and the Sustainable Management of Forests*]. https://colaboracion.dnp.gov.co/CDT/Conpes/Econ%C3%B3micos/4021.pdf.

CONPES 4050. 2021. CONPES 4050. *Poítica para la consolidación del Sistema Nacional de Áreas Protegidas.* [*Policy for the consolidation of the National System of Protected Areas*]. SINAP. https://colaboracion.dnp.gov.co/CDT/Conpes/Econ%C3%B3micos/4050.pdf.

CONPES 4062. 2021. CONPES 4062. *Politica Nacional de Propiedad Intelectual.* [*National Intellectual Property Policy*]. https://colaboracion.dnp.gov.co/CDT/Conpes/Econ%C3%B3micos/4062.pdf.

CONPES 4069. 2021. CONPES 4069. *Política nacional de ciencia, tecnología e innovación 2022-2031.* [*National science, technology and innovation policy 2022-2031*]. https://colaboracion.dnp.gov.co/CDT/Conpes/Econ%C3%B3micos/4069.pdf.

CONPES 4075. 2022. CONPES 4075. *Política de Transición Energética.* [*Energy Transition Policy*]. https://colaboracion.dnp.gov.co/CDT/Conpes/Econ%C3%B3micos/4075.pdf.

CONPES 4085. 2022. Conpes 4085. *Política de internacionalización para el desarrollo productivo regional.* [*Internationalization policy for regional productive development*]. https://colaboracion.dnp.gov.co/CDT/Conpes/Econ%C3%B3micos/4085.pdf.

CONPES 4098. 2022. CONPES 4098. *Política para impulsar la competitividad agropecuaria.* [*Policy to promote agricultural competitiveness*]. https://colaboracion.dnp.gov.co/CDT/Conpes/Econ%C3%B3micos/4098.pdf.

Consejo Económico para América Latina (CEPAL). 2015. *Bioeconomia, nuevas oportunidades para la agricultura.* [*Bioeconomy, new opportunities for agriculture*].

Conteratto, C., F. D. Artuzo, O. I. Benedetti Santos, and E. Talamini. 2021. "Biorefinery: A comprehensive concept for the sociotechnical transition toward bioeconomy." *Renewable and Sustainable Energy Reviews* 151. doi: 10.1016/j.rser.2021.111527.

Corrales-Ureña, Y. R., C. Villalobos-Bermúdez, R. Pereira, M. Camacho, E. Estrada, O. Argüello-Miranda, and J. R. Vega-Baudrit. 2018. "Biogenic silica-based microparticles obtained as a sub-product of the nanocellulose extraction process from pineapple peels." *Sci Rep* 8 (1):10417. doi: 10.1038/s41598-018-28444-4.

Costa, Jose Augusto Martini. 2020. *From Sugarcane To Ethanol: The Historical Process That Transformed Brazil Into A Biofuel Superpower*. Vol. 10, *2020*. Biofuel; Brazil; Historical Institutionalism; Sugarcane.

Daza Serna, L. V., J. C. Solarte Toro, S. Serna Loaiza, Y. Chacón Perez, and C. A. Cardona Alzate. 2016. "Agricultural Waste Management Through Energy Producing Biorefineries: The Colombian Case." *Waste and Biomass Valorization* 7 (4):789-798. doi: 10.1007/s12649-016-9576-3.

de Arruda, Priscila Vaz, Rita de Cássia Lacerda Brambilla Rodrigu, Débora Danielle Virgínio da Silva, and Maria das Graças de Almeida Felipe. 2011. "Evaluation of hexose and pentose in pre-cultivation of Candida guilliermondii on the key enzymes for xylitol production in sugarcane hemicellulosic hydrolysate." *Biodegradation* 22:815-822. doi: 10.1007/s10532-010-9397-1.

Del Pozo, C., J. Bartrolí, S. Alier, N. Puy, and E. Fàbregas. 2020. "Production of antioxidants and other value-added compounds from coffee silverskin via pyrolysis under a biorefinery approach." *Waste Manag* 109:19-27. doi: 10.1016/j.wasman. 2020.04.044.

Departmento Nacional de Planeación [National Planning Department], DNP. 2018. *Bioeconomía* [*Bioeconomy*]. https://www.dnp.gov.co/Crecimiento-Verde/Ejes-estrategicos/Paginas/Bioeconom%C3%ADa.aspx. Bogotá: DNP.

Dutta, Shanta, Iris K.M. Yu, Daniel C.W. Tsang, Yun Hau Ng, Yong Sik Ok, James Sherwood, and James H. Clark. 2019. "Green synthesis of gamma-valerolactone (GVL) through hydrogenation of biomass-derived levulinic acid using non-noble metal catalysts: A critical review." *Chemical Engineering Journal* 372:992-1006. doi: 10.1016/j.cej.2019.04.199.

Dutta, Shanta, Qiaozhi Zhang, Yang Cao, Chunfei Wu, Konstantinos Moustakas, Shicheng Zhang, Ka-Hing Wong, and Daniel C. W. Tsang. 2022. "Catalytic valorisation of various paper wastes into levulinic acid, hydroxymethylfurfural, and furfural: Influence of feedstock properties and ferric chloride." *Bioresource Technology* 357:127376. doi: 10.1016/j.biortech.2022.127376.

Dzyazko, Yuliya S., Olexii V. Palchik, Vladimir M. Ogenko, Leon Y. Shtemberg, Valerii I. Bogomaz, Sergii A. Protsenko, Vladimir G. Khomenko, Irina S. Makeeva, Oxana V. Chernysh, and Olexander G. Dzyazko. 2019. "Nanoporous Biochar for Removal of Toxic Organic Compounds from Water." *Nanophotonics, Nanooptics, Nanobiotechnology, and Their Applications, Cham,* 2019//.

E4tech (UK) Ltd. 2017. *Advanced drop-in biofuels UK production capacity outlook to 2030.* Edited by Temple and Scarlett Research. for Department for Transport in partnership with TRL. UK: Department for Transport.

Eixenberger, Daniela, Ana-Francis Carballo-Arce, José-Roberto Vega-Baudrit, Humberto Trimino-Vazquez, Luis Roberto Villegas-Peñaranda, Anne Stöbener, Francisco Aguilar, Jose-Aníbal Mora-Villalobos, Manuel Sandoval-Barrantes, Paul Bubenheim, and Andreas Liese. 2022. "Tropical agroindustrial biowaste revalorization through integrative biorefineries—review part II: pineapple, sugarcane and banana by-products in Costa Rica." *Biomass Conversion and Biorefinery*. doi: 10.1007/s13399-022-02721-9.

European Comission. 2015. From the sugar platform to biofuels and biochemicals. In *Final report for the European Commission Directorate General Energy*. Bruselas.

European Commission's Research DG. 2005. "*New Perspectives on the knowledge based bio-economy*." Transforming life science knowledge into new, sustainable, eco-efficient and competitive products.

Fang, Z., R. Smith, and X. Qui. 2017. "Production of Platform Chemicals from Sustainable Resources." *Production of Platform Chemicals from Sustainable Resources*: 375-410.

Felipe Hernández-Pérez, Andrés, Priscila Vaz de Arruda, Luciane Sene, Silvio Silvério da Silva, Anuj Kumar Chandel, and Maria das Graças de Almeida Felipe. 2019. "Xylitol bioproduction: state-of-the-art, industrial paradigm shift, and opportunities for integrated biorefineries." *Critical Reviews in Biotechnology* 39:924-943. doi: 10.1080/07388551.2019.1640658.

Franco-Bacca, Adriana Paola, Fernando Cervantes-Alvarez, Juan Daniel Macías, Joan Alexis Castro-Betancur, Reynell Junior Pérez-Blanco, Oscar Hernán Giraldo Osorio, Nayda Patricia Arias Duque, Geonel Rodríguez-Gattorno, and Juan José Alvarado-Gil. 2021. "Heat Transfer in Cassava Starch Biopolymers: Effect of the Addition of Borax." *Polymers* 13 (23):4106.

Gandla, Madhavi, Carlos Martín, and Leif Jönsson. 2018. "Analytical Enzymatic Saccharification of Lignocellulosic Biomass for Conversion to Biofuels and Bio-Based Chemicals." *Energies* 11:2936. doi: 10.3390/en11112936.

García Prieto, Carla V. 2018. "*Diseño óptimo de biorrefinerías integradas basadas en microalgas.*" Doctor en Ingeniería Química Doctoral, Ingeniería Química, Universidad Nacional del Sur. [*Optimal design of integrated biorefineries based on microalgae.* Ph.D. in Chemical Engineering Ph.D., Chemical Engineering, Universidad Nacional del Sur.].

Georgescu-Roegen, Nicholas. 1975. "Energy and Economic Myths." *Southern Economic Journal* 41 (3):347-381.

German Federal Goverment. 2021. *Bioeconomy in Germany - Background.* https://biooekonomie.de/en/bioeconomy-germany-background#:~:text=The%20German%20Federal%20Government%20defines,of%20a%20future%2Doriented%20economy. accesed july 25, 2022. Germany.

Giosuè, Chiara, Alessandra Mobili, Giuseppe Toscano, Maria Letizia Ruello, and Francesca Tittarelli. 2016. "Effect of Biomass Waste Materials as Unconventional Aggregates in Multifunctional Mortars for Indoor Application." *Procedia Engineering* 161:655-659. doi: 10.1016/j.proeng.2016.08.724.

Giraldo, Jhonny Alejandro Poveda, Mariana Ortiz Sanchez, Juan Camilo Solarte-Toro, and Carlos Ariel Cardona Alzate. 2020. "8 - Economic and social aspects of biorefineries." In *Recent Advances in Bioconversion of Lignocellulose to Biofuels and Value-Added Chemicals within the Biorefinery Concept*, edited by Edivaldo Ximenes Ferreira Filho, Leonora Rios de Souza Moreira, Eduardo de Aquino Ximenes and Cristiane Sanchez Farinas, 199-231. Elsevier.

Gobierno de Uruguay. [Government of Uruguay]. 2019. Ley N° 19829. [Law No. 19829]. *Gestión Integral de Residuos.* [*Comprehensive Waste Management*]. https://www.impo.com.uy/bases/leyes/19829-2019.

Gobierno de Uruguay. [Government of Uruguay]. 2021. *Plan Nacional de Gestión de Residuos.* [*National Waste Management Plan*]. https://www.ambiente.gub.uy/oan/documentos/PNGR-general.pdf.

Gómez-Cruz, Irene, María del Mar Contreras, Inmaculada Romero, and Eulogio Castro. 2021. "A biorefinery approach to obtain antioxidants, lignin and sugars from exhausted olive pomace." *Journal of Industrial and Engineering Chemistry* 96:356-363. doi: https://doi.org/10.1016/j.jiec.2021.01.042.

Grangeia, Carolina, Luan Santos, and Lira Luz Benites Lazaro. 2022. "The Brazilian biofuel policy (RenovaBio) and its uncertainties: An assessment of technical, socioeconomic and institutional aspects." *Energy Conversion and Management: X* 13:100156. doi: https://doi.org/10.1016/j.ecmx.2021.100156.

Hbaieb, S., W. Kammoun, C. Delaite, M. Abid, S. Abid, and R. El Gharbi. 2015. "New Copolyesters Containing Aliphatic and Bio-Based Furanic Units by Bulk Copolycondensation." *Journal of Macromolecular Science, Part A* 52:365-373. doi: 10.1080/10601325.2015.1018807.

Hernández-Pérez, Andrés Felipe, Priscila Vaz de Arruda, and Maria das Graças de Almeida Felipe. 2016. "Sugarcane straw as a feedstock for xylitol production by Candida guilliermondii FTI 20037." *Brazilian Journal of Microbiology* 47:489-496. doi: 10.1016/j.bjm.2016.01.019.

Hong, Kuk-Ki, and Jens Nielsen. 2012. "Metabolic engineering of Saccharomyces cerevisiae: a key cell factory platform for future biorefineries." *Cellular and Molecular Life Sciences* 69:2671-2690. doi: 10.1007/s00018-012-0945-1.

Hughes, Stephen R., Juan Carlos López-Núñez, Marjorie A. Jones, Bryan R. Moser, Elby J. Cox, Mitch Lindquist, Luz Ángela Galindo-Leva, Néstor M. Riaño-Herrera, Nelson Rodriguez-Valencia, Fernando Gast, David L. Cedeño, Ken Tasaki, Robert C. Brown, Al Darzins, and Lane Brunner. 2014. "Sustainable conversion of coffee and other crop wastes to biofuels and bioproducts using coupled biochemical and thermochemical processes in a multi-stage biorefinery concept." *Applied Microbiology and Biotechnology* 98:8413-8431. doi: 10.1007/s00253-014-5991-1.

Hugo Chavarría, Eduardo Trigo (IICA), and and Adrián Rodríguez (ECLAC). 2019. *"The bioeconomy: a catalyst for the sustainable development of agriculture and rural territories in LAC* http://www.iica-ecuador.org/sisbio/doc_informacion/IICA_Cap4_Eng_V4.pdf. accesed july 2022."

IEA Bioenergy. 2009. Biorefineries: Adding value to the sustainable utilisation of biomass. In *Energy Agency.*

IEA Bioenergy. 2021. *"Renewable Biofuels."* accessed July 25. https://www.iea.org/reports/renewables-2021/biofuels?mode=transport&region=World&publication=2021&flow=Consumption&product=Ethanol.

IEA Bioenergy., Agenzia nazionale per le nuove tecnologie l'energia e lo sviluppo economico sostenibile., and ENEA. [National Agency for New Energy Technology and Sustainable Economic Development, and ENEA]. 2022. *Biorefinery Plant Portal. IEA bioenergy,* Task 42.

IICA. 2020. *Bioeconomía: Una estrategia de desarrollo para la argentina del siglo xxi Impulsando a la bioeconomía como modelo de desarrollo sustentable: entre las políticas públicas y las estrategias privadas* [*Bioeconomy: A development strategy for Argentina in the 21st century Promoting the bioeconomy as a model of sustainable development: between public policies and private strategies*]. https://repositorio.iica.int/bitstream/handle/11324/12478/BVE20108164e.pdf?sequence=1&isAllowed=y accesed july 2022.

International Furan Chemicals B. V. 2016. *Historical overview and industrial development of furfural.*

Isikgor, Furkan H., and C. Remzi Becer. 2015. "Lignocellulosic biomass: a sustainable platform for the production of bio-based chemicals and polymers." *Polymer Chemistry* 6:4497-4559. doi: 10.1039/C5PY00263J.

Jofre, Fanny Machado, Fernanda Weber Bordini, Italo de Andrade Bianchini, Sarah de Souza Queiroz, Tatiane da Silva Boaes, Andrés Felipe Hernández-Pérez, and Maria das Graças de Almeida Felipe. 2022. "8 - Xylitol and sorbitol: production routes, challenges and opportunities in biorefineries integration." In *Production of Top 12 Biochemicals Selected by USDOE from Renewable Resources*, edited by Anuj K. Chandel and Fernando Segato, 233-268. Elsevier.

Johnson, Thomas G., and Ira Altman. 2014. "Rural development opportunities in the bioeconomy." *Biomass and Bioenergy* 63:341-344. doi: https://doi.org/10.1016/j.biombioe.2014.01.028.

José Roberto Vega Baudrit. 2007. "La Nanotecnología en Costa Rica: la experiencia en el LANOTEC [Nanotechnology in Costa Rica: the experience at LANOTEC]. https://www.cientec.or.cr/exploraciones/ponencias2007/JoseRobertoVega.pdf." *IX Congreso Nacional de Ciencias Exploraciones fuera y dentro* del aula 24 y 25 de agosto, 2007, Instituto Tecnológico de Costa Rica Cartago, Costa Rica. [*IX National Congress of Sciences Explorations outside and inside the classroom* August 24 and 25, 2007, Technological Institute of Costa Rica Cartago, Costa Rica].

Juan-García, A., C. Gallego, and G. Font. 2015. "Toxicidad del Bisfenol A: Revisión [Bisphenol A Toxicity: Review]." *Revista de Toxicologia* 32:144-160.

Kalantari, Mohammad, Mohammad Kazemeini, and Ayyoob Arpanaei. 2013. "Evaluation of biodiesel production using lipase immobilized on magnetic silica nanocomposite particles of various structures." *Biochemical Engineering Journal* 79:267-273. doi: https://doi.org/10.1016/j.bej.2013.09.001.

Kamm, Birgit. 2014. "Biorefineries – their scenarios and challenges." *Pure and Applied Chemistry* 86 (5):821-831. doi: doi:10.1515/pac-2013-1035.

Kapanji, Kutemba K., Kathleen F. Haigh, and Johann F. Görgens. 2019. "Techno-economic analysis of chemically catalysed lignocellulose biorefineries at a typical sugar mill: Sorbitol or glucaric acid and electricity co-production." *Bioresource Technology* 289:121635. doi: https://doi.org/10.1016/j.biortech.2019.121635.

Khoshnevisan, Benyamin, Meisam Tabatabaei, Panagiotis Tsapekos, Shahin Rafiee, Mortaza Aghbashlo, Susanne Lindeneg, and Irini Angelidaki. 2020. "Environmental life cycle assessment of different biorefinery platforms valorizing municipal solid waste to bioenergy, microbial protein, lactic and succinic acid." *Renewable and Sustainable Energy Reviews* 117. doi: 10.1016/J.RSER.2019.109493.

King, David. 2010. *The future of industrial biorefineries.* Cologny, Geneva: World Economic Forum.

Kirchherr, Julian, Denise Reike, and Marko Hekkert. 2017. "Conceptualizing the circular economy: An analysis of 114 definitions." *Resources, Conservation and Recycling* 127:221-232. doi: 10.1016/j.resconrec.2017.09.005.

Kogje, Anushree B, and Anand Ghosalkar. 2017. "Xylitol production by genetically modified industrial strain of Saccharomyces cerevisiae using glycerol as co-substrate." *Journal of Industrial Microbiology and Biotechnology* 44:961-971. doi: 10.1007/s10295-017-1914-3.

Koutinas, Apostolis A., Anestis Vlysidis, Daniel Pleissner, Nikolaos Kopsahelis, Isabel Lopez Garcia, Ioannis K. Kookos, Seraphim Papanikolaou, Tsz Him Kwan, and Carol Sze Ki Lin. 2014. "Valorization of industrial waste and by-product streams via fermentation for the production of chemicals and biopolymers." *Chemical Society Reviews* 43:2587-2627. doi: 10.1039/C3CS60293A.

Lei, Tingzhou, Zhiwei Wang, Xia Chang, Lu Lin, Xiaoyu Yan, Yincong Sun, Xinguang Shi, Xiaofeng He, and Jinling Zhu. 2016. "Performance and emission characteristics of a diesel engine running on optimized ethyl levulinate–biodiesel–diesel blends." *Energy* 95:29-40. doi: 10.1016/j.energy.2015.11.059.

Li, Jingen, Liangcai Lin, Tao Sun, Jing Xu, Jingxiao Ji, Qian Liu, and Chaoguang Tian. 2020. Direct production of commodity chemicals from lignocellulose using *Myceliophthora thermophila*. In *Metabolic Engineering*: Elsevier Inc.

Li, Xiaodan, Pei Jia, and Tiefeng Wang. 2016. "Furfural: A Promising Platform Compound for Sustainable Production of C4 and C5 Chemicals." *ACS Catalysis* 6 (11):7621-7640. doi: 10.1021/acscatal.6b01838.

Liu, Changshui, Yamei Ding, Mo Xian, Min Liu, Huizhou Liu, Qingjun Ma, and Guang Zhao. 2017. "Malonyl-CoA pathway: a promising route for 3-hydroxypropionate biosynthesis." *Critical Reviews in Biotechnology* 37:933-941. doi: 10.1080/07388551.2016.1272093.

Magdalena, Borges, Atilio Deana, Lucia Pittaluga, Carolina Balian, and Adrián Rodríguez. 2021. *Contribución de la bioeconomía a la recuperación pospandemia de COVID-19 en el Uruguay Biotecnología y valorización de subproductos agropecuarios y agroindustriales*. [*Contribution of the bioeconomy to the post-COVID-19 pandemic recovery in Uruguay Biotechnology and recovery of agricultural and agro-industrial by-products*].

Mao, Feng, Shuainan Chen, Qiao Zhang, Long Yang, Feifei Wan, Dabo Jiang, Manman Xiong, Chao Zhang, Yachun Liu, and Zaihui Fu. 2020. "Ru/P-Containing Porous Biochar-Efficiently Catalyzed Cascade Conversion of Cellulose to Sorbitol in Water under Medium-Pressure H2 Atmosphere." *Bulletin of the Chemical Society of Japan* 93 (8):1026-1035. doi: 10.1246/bcsj.20200095.

Marín-Valencia, Paula Andrea, Estefanny Carmona-Garcia, Jhonny Alejandro Poveda Giraldo, Nayda Patricia Arias Duque, Lina Fernanda Ballesteros, and Carlos Ariel Cardona Alzate. 2021. "The integral use of aromatic plants: prefeasibility comparison of stand-alone and biorefinery processes using thyme (Thymus vulgaris) as base case." *Biomass Conversion and Biorefinery* 11 (2):681-691. doi: 10.1007/s13399-020-00734-w.

Market Research Future. 2020. Global Isoprene Market: Information by Grade (Polymer Grade and Chemical Grade). In : *Application (Styrene Isoprene Styrene, Isobutylene Isoprene, Polyisoprene, Block Co-Polymer, and Others), End-Use Industry (Automotive, Medical, Construction, Personal Care, Sports & Leisure, and Others), and Region (North America, Europe, Asia-Pacific*.

Master Watch. 2022. *"2,5-Furandicarboxylic Acid (FDCA) Market Size from 2022 to 2028."* accessed july 8. https://www.marketwatch.com/press-release/25-furandi carboxylic-acid-fdca-market-size-from-2022-to-2028-2022-06-09.

Mazzoli, Roberto. 2020. "Metabolic engineering strategies for consolidated production of lactic acid from lignocellulosic biomass." *Biotechnology and Applied Biochemistry* 67:61-72. doi: 10.1002/BAB.1869.

Mika, László T., Edit Cséfalvay, and Áron Németh. 2018. "Catalytic Conversion of Carbohydrates to Initial Platform Chemicals: Chemistry and Sustainability." *Chemical Reviews* 118:505-613. doi: 10.1021/ACS.CHEMREV.7B00395.

Ministerio de Ambiente y Desarrollo Sostenible. 1996. *Política Nacional de Biodiversidad.* [Ministry of Environment and Sustainable Development. 1996. *National Biodiversity Policy].*

Ministerio de Ambiente y Desarrollo Sostenible. 2012. *Política para la Gestión Integral de la Biodiversidad y los Servicios Ecosistémicos.* [Ministry of Environment and Sustainable Development. 2012. *Policy for the Comprehensive Management of Biodiversity and Ecosystem Services*]. https://www.minambiente.gov.co/wp-content/uploads/2021/10/Poli%CC%81tica-Nacional-de-Gestio%CC%81n-Integral-de-la-Biodiver.pdf.

Ministerio de Ciencia Tecnología e Innovación. 2020. *Bioeconomía para una potencia viva y diversa; Hacia una sociedad impulsada por el conocimiento.* [Ministry of Science Technology and Innovation. 2020. *Bioeconomy for a living and diverse power; Towards a knowledge-driven society*]. https://minciencias.gov.co/sites/default/files/upload/paginas/bioeconomia_para_un_crecimiento_sostenible-qm_print.pdf.

Ministerio del Medio Ambiente de Chile. 2020. *Estrategia nacional de residuos orgánicos Chile 2040.* [Ministry of the Environment of Chile. 2020. *Chile 2040 National Organic Waste Strategy.].* https://economiacircular.mma.gob.cl/wp-content/uploads/2021/03/Estrategia-Nacional-de-Residuos-Organicos-Chile-2040.pdf.

Mohan, D., K. Abhishek, A. Sarswat, M. Patel, P. Singh, and C. Pittman. 2018. "Biochar production and applications in soil fertility and carbon sequestration – a sustainable solution to crop-residue burning in India." *RSC Advances* 8(1):508-520. doi: https://doi.org/10.1039/c7ra10353k.

Moncada B, Jonathan, Valentina Aristizábal M, and Carlos A. Cardona A. 2016. "Design strategies for sustainable biorefineries." *Biochemical Engineering Journal* 116:122-134. doi: https://doi.org/10.1016/j.bej.2016.06.009.

Morakile, Tumelo, Mohsen Mandegari, Somayeh Farzad, and Johann.F. Görgens. 2022. "Comparative techno-economic assessment of sugarcane biorefineries producing glutamic acid, levulinic acid and xylitol from sugarcane." *Industrial Crops and Products* 184:115053. doi: 10.1016/j.indcrop.2022.115053.

Mora-Sandí, Anthony, Abigail Ramírez-González, Luis Castillo-Henríquez, Mary Lopretti-Correa, and José Roberto Vega-Baudrit. 2021. "Persea Americana Agro-Industrial Waste Biorefinery for Sustainable High-Value-Added Products." *Polymers* 13 (11):1727.

Moreda, Iván López. 2016. "The potential of biogas production in Uruguay." *Renewable and Sustainable Energy Reviews* 54:1580-1591. doi: https://doi.org/10.1016/j.rser.2015.10.099.

Murrell, J. Colin, Terry J. McGenity, and Andrew T. Crombie. 2020. "Microbial metabolism of isoprene: a much-neglected climate-active gas." *Microbiology* 166:600-613. doi: 10.1099/mic.0.000931.

Nawaj Alam, Shahrukh, Bhaskar Singh, and Abhishek Guldhe. 2021. "Aquatic weed as a biorefinery resource for biofuels and value-added products: Challenges and recent advancements." *Cleaner Engineering and Technology* 4:100235. doi: https://doi.org/10.1016/j.clet.2021.100235.

Nghiem, N., C. Nguyen, C. Drapcho, and T. Walker. 2013. "Sweet sorghum biorefinery for production of fuel ethanol and value-added co-products." *Biological Engineering Transactions* 6. doi: 10.13031/bet.6.9926.

Nunes, Bruno. 2017. *Brazilian Policies and Programs in Bioeconomy.* https://fapesp.br/eventos/2017/6dialogue/08-11-17/15h45_Bruno_Nunes.pdf. accessed on july 2022.

OCDE. 2009. *The Bioeconomy to 2030: Designing a policy agenda*, vol. 9789264056. Organisation for Economic Cooperation and Development (OECD); 2009. https://doi.org/10.1787/9789264056886-en.

ONU. 1992. *Convenio de Diversidad Biológica.* [*Biological Diversity Convention*]. http://chmcolombia.co/politica-y-cooperacion/acuerdos/cdb/#:~:text=El%20Convenio%20de%20la%20Diversidad,el%20Medio%20Ambiente%20y%20Desarrollo.

Orejuela Escobar, Lourdes M. 2018. "Biorefinería: un modelo de negocios de productos de alto valor agregado a partir de desechos agrícolas e industriales y promotora de desarrollo sustentable en el contexto de la bioeconomía." [Biorefinery: a business model for high value-added products from agricultural and industrial waste and a promoter of sustainable development in the context of the bioeconomy]. *Memorias Y Boletines De La Universidad Del Azuay* [*Proceedings and Bulletins of the University of Azuay*] 1:199-214.

Ortiz-Sanchez, M., J. C. Solarte-Toro, C. E. Orrego-Alzate, C. D. Acosta-Medina, and C. A. Cardona-Alzate. 2021. "Integral use of orange peel waste through the biorefinery concept: an experimental, technical, energy, and economic assessment." *Biomass Conversion and Biorefinery* 11 (2):645-659. doi: 10.1007/s13399-020-00627-y.

Ortiz-Sanchez, Mariana, and Carlos Ariel Cardona Alzate. 2022. "Analysis of the routes for biomass processing towards sustainable development in the conceptual design step: Strategy based on the compendium of bioprocesses portfolio." *Bioresource Technology* 350:126852. doi: https://doi.org/10.1016/j.biortech.2022.126852.

Pacheco-Torgal, F. 2020. "1 - Introduction to biobased materials and biotechnologies for eco-efficient construction." In *Bio-Based Materials and Biotechnologies for Eco-Efficient Construction*, edited by Fernando Pacheco-Torgal, Volodymyr Ivanov and Daniel C. W. Tsang, 1-16. Woodhead Publishing.

Padi, Richard Kingsley, and Annie Chimphango. 2020. "Feasibility of commercial waste biorefineries for cassava starch industries: Techno-economic assessment." *Bioresource Technology* 297:122461. doi: https://doi.org/10.1016/j.biortech.2019.122461.

Papageorgiou, G. Z., V. Tsanaktsis, D. G. Papageorgiou, S. Exarhopoulos, M. Papageorgiou, and D. N. Bikiaris. 2014. "Evaluation of polyesters from renewable resources as alternatives to the current fossil-based polymers. Phase transitions of poly(butylene 2,5-furan-dicarboxylate)." *Polymer* 55:3846-3858. doi: 10.1016/j.polymer.2014.06.025.

Papageorgiou, George Z., Dimitrios G. Papageorgiou, Zoi. Terzopoulou, and Dimitrios N. Bikiaris. 2016. Production of bio-based 2,5-furan dicarboxylate polyesters: Recent progress and critical aspects in their synthesis and thermal properties. In *European Polymer Journal*.

Parra-Ramírez, Daniela, Juan Camilo Solarte-Toro, and Carlos Ariel Cardona-Alzate. 2020. "Techno-Economic and Environmental Analysis of Biogas Production from Plantain Pseudostem Waste in Colombia." *Waste and Biomass Valorization* 11 (7):3161-3171. doi: 10.1007/s12649-019-00643-8.

Patermann, Christian, and Alfredo Aguilar. 2018. "The origins of the bioeconomy in the European Union." *New Biotechnology* 40:20-24. doi: 10.1016/j.nbt.2017.04.002.

Patrizi, Nicoletta, Morena Bruno, Fabrizio Saladini, Maria Laura Parisi, Riccardo M. Pulselli, Anne Belinda Bjerre, and Simone Bastianoni. 2020. "Sustainability Assessment of Biorefinery Systems Based on Two Food Residues in Africa." *Frontiers in Sustainable Food Systems* 4. doi: 10.3389/fsufs.2020.522614.

Peña, S., and J. Lopez. 2020. "Sustainable development and learning opportunity of biorefineries: A biomass alternative." *Revista de ciencias sociales* 26:401-413.

Peng, Lincai, Lu Lin, Junhua Zhang, Junping Zhuang, Beixiao Zhang, and Yan Gong. 2010. "Catalytic Conversion of Cellulose to Levulinic Acid by Metal Chlorides." *Molecules (Basel, Switzerland)* 15:5258-72. doi: 10.3390/molecules15085258.

Perez., Andrea Teresa Espinoza. 2017. *"Biorefinery supply chain design optimization under sustainability dimensions."* Doctoral, Computational Engineering, Finance, and Science [cs.CE], University of Lorraine (ffNNT: 2017LORR0222ff. fftel-0181626).

Pileidis, Fi., and M. Titirici. 2016a. Levulinic Acid Biorefineries: New Challenges for Efficient Utilization of Biomass. In *Chem Sus Chem*.

Pileidis, Filoklis D., and Maria-Magdalena Titirici. 2016b. "Levulinic Acid Biorefineries: New Challenges for Efficient Utilization of Biomass." *Chem Sus Chem* 9 (6):562-582. doi: https://doi.org/10.1002/cssc.201501405.

Potrč, Sanja, Lidija Čuček, Mariano Martin, and Zdravko Kravanja. 2020. "Synthesis of European Union Biorefinery Supply Networks Considering Sustainability Objectives." *Processes* 8 (12):1588.

Prabha, Syama, Aravind K. Vijay, Rony Rajan Paul, and Basil George. 2022. "Cyanobacterial biorefinery: Towards economic feasibility through the maximum valorization of biomass." *Science of The Total Environment* 814:152795. doi: 10.1016/j.scitotenv.2021.152795.

Queiroz, Sarah S., Fanny M. Jofre, Solange I. Mussatto, and Maria das Graças A. Felipe. 2022. "Scaling up xylitol bioproduction: Challenges to achieve a profitable bioprocess." *Renewable and Sustainable Energy Reviews* 154:111789. doi: 10.1016/j.rser.2021.111789.

Rajabinejad, H., I. Bucişcanu, and S. Maier. 2019. Current approaches for raw wool waste management and unconventional valorization: A review. In *Environmental Engineering and Management Journal*.

Rajesh, R. O., T. K. Godan, A. K. Rai, D. Sahoo, A. Pandey, and P. Binod. 2019. "Biosynthesis of 2,5-furan dicarboxylic acid by Aspergillus flavus APLS-1: Process optimization and intermediate product analysis." *Bioresource Technology* 284:155-160. doi: 10.1016/j.biortech.2019.03.105.

Ramcilovic-Suominen, S., and H. Pülzl. 2018. "Sustainable development – A 'selling point' of the emerging EU bioeconomy policy framework?" *Journal of Cleaner Production* 172:4170-4180. doi: https://doi.org/10.1016/j.jclepro.2016.12.157.

Ratan, Jatinder Kumar, Manjeet Kaur, and Bharadwaj Adiraju. 2018. "Synthesis of activated carbon from agricultural waste using a simple method: Characterization, parametric and isotherms study." *Materials Today: Proceedings* 5:3334-3345. doi: 10.1016/j.matpr.2017.11.576.

Redondo-Gómez, Carlos, Maricruz Rodríguez Quesada, Silvia Vallejo Astúa, José Pablo Murillo Zamora, Mary Lopretti, and José Roberto Vega-Baudrit. 2020. "Biorefinery of Biomass of Agro-Industrial Banana Waste to Obtain High-Value Biopolymers." *Molecules* 25 (17):3829.

Rengel, Rocío, Inmaculada Giraldez, Manuel J. Díaz, Trinidad García, Javier Vigara, and Rosa León. 2022. "Simultaneous production of carotenoids and chemical building blocks precursors from chlorophyta microalgae." *Bioresource Technology* 351:127035. doi: https://doi.org/10.1016/j.biortech.2022.127035.

Rodríguez, A. G., O. Sotomayor, and M. Rodrigues. 2019. *"Hacia una bioeconomía sostenible en América Latina y el Caribe: elementos para una visión regional."* [*Towards a sustainable bioeconomy in Latin America and the Caribbean: elements for a regional vision*]. Serie Recursos Naturales y Desarrollo, [Natural Resources and Development Series], Santiago.

Rodríguez, Adrián G., Andrés O. Mondaini, and Maureen A. Hitschfeld. 2017. *Bioeconomía en América Latina y el Caribe Contexto global y regional y perspectivas, Desarrollo Productivo*: CEPAL-ONU. [*Bioeconomy in Latin America and the Caribbean Global and regional context and perspectives, Productive Development*].

Rodríguez-Aguilera, Jessica., Adolfo. Brown-Gómez, Amaury. Delgado, and Georgina. Michelena. 2020. "Producción de ácido levulínico: una revisión bibliográfica [Levulinic acid production: a literature review]." *ICIDCA sobre los derivados de la caña de azúcar* 52:36-46.

Rodriguez-Fernandez, Silvia, Ismael Díaz, María González-Miquel, Emilio J. González, and Manuel Rodriguez. 2022. "Optimization-based analysis of integrated lignocellulosic biorefineries in Spain focusing on building blocks." *Biomass Conversion and Biorefinery*. doi: 10.1007/s13399-021-02247-6.

Rorrer, N. A., D. R. Vardon, J. R. Dorgan, E. J. Gjersing, and G. T. Beckham. 2017. "Biomass-derived monomers for performance-differentiated fiber reinforced polymer composites." *Green Chemistry* 19:2812-2825. doi: 10.1039/c7gc00320j.

Sacramento-Rivero, J. C., G. Romero, E. Cortés-Rodríguez, E. Pech, and S. Blanco-Rosete. 2010. A diagnostic study on the development of biorefineries in Mexico. *Rev. Mex. Ing. Quím* [online]. 2010, vol.9, n.3, pp.261-283.

Sajid, M., X. Zhao, and D. Liu. 2018. "Production of 2,5-furandicarboxylic acid (FDCA) from 5-hydroxymethylfurfural (HMF): Recent progress focusing on the chemical-catalytic routes." *Green chem* 24:5427-5453.

Santiago, B.L.S., and R. Guirardello. 2020. "5-hydroxymethylfurfural production in a lignocellulosic biorefinery: Techno-economic analysis." *Chemical Engineering Transactions* 80:139-144. doi: 10.3303/CET2080024.

Sara Piedrahita Rodriguez. 2020. *"Design of biorefineries with multiple raw materials for the use of agroindustrial waste in post-conflict zones in colombia." Magister Magister, Chemical Enginering, Universidad Nacional de Colombia. Sede Manizales.* [Magister

Magister, Chemical Engineering, National University of Colombia. Manizales Headquarters].

Sari, Yessie W., Sumaya Yulia Putri, Noor Intan, Abdurrahman Bahtiar, and Mersi Kurniati. 2021. "The effect of sorbitol and sweet sorghum to carrageenan ratio on the physicochemical properties of sweet sorghum/carrageenan bioplastics." *Biomass Conversion and Biorefinery*. doi: 10.1007/s13399-020-01254-3.

Sauer, M., D. Porro, D. Mattanovich, and P. Branduardi. 2008. "Microbial production of organic acids: expanding the markets." *Trends Biotechnol* 26 (2):100-8. doi: 10.1016/j.tibtech.2007.11.006.

Schmidt, Stanislaus, Felix J. Gatti, Manuel Luitz, Benjamin S. Ritter, Bernd Bruchmann, and Rolf Mülhaupt. 2017. "Erythritol Dicarbonate as Intermediate for Solvent- and Isocyanate-Free Tailoring of Bio-Based Polyhydroxyurethane Thermoplastics and Thermoplastic Elastomers." *Macromolecules* 50:2296-2303. doi: 10.1021/acs.macromol.6b02787.

Senthil, Chenrayan, and Chang Woo Lee. 2021. "Biomass-derived biochar materials as sustainable energy sources for electrochemical energy storage devices." *Renewable and Sustainable Energy Reviews* 137:110464. doi: https://doi.org/10.1016/j.rser.2020.110464.

Servian-Rivas, Luis David, Elia Ruiz Pachón, Manuel Rodríguez, María González-Miquel, Emilio J. González, and Ismael Díaz. 2022. "Techno-economic and environmental impact assessment of an olive tree pruning waste multiproduct biorefinery." *Food and Bioproducts Processing* 134:95-108. doi: 10.1016/j.fbp.2022.05.003.

Shahid, Muhammad Kashif, Ayesha Batool, Ayesha Kashif, Muhammad Haq Nawaz, Muhammad Aslam, Nafees Iqbal, and Younggyun Choi. 2021. "Biofuels and biorefineries: Development, application and future perspectives emphasizing the environmental and economic aspects." *Journal of Environmental Management* 297:113268. doi: https://doi.org/10.1016/j.jenvman.2021.113268.

Sharma, A., A. Sharma, J. Singh, P. Sharma, G. S. Tomar, S. Singh, and L. Nain. 2020. "A biorefinery approach for the production of ferulic acid from agroresidues through ferulic acid esterase of lactic acid bacteria." *3 Biotech* 10 (8):367. doi: 10.1007/s13205-020-02360-9.

Sivamani, S. Chandrasekaran, A., M. Balajii, and A. Hosseini-Bandegharaei Shanmugaprakash, A. Baskar, R. 2018. "Evaluation of the potential of cassava-based residues for biofuels production." *Environ. Sci. Biotechnol* 17:553-570. doi: https://doi.org/10.1007/s11157-018-9475-0.

Solarte-Toro, J. C., and C. A. Cardona Alzate. 2021a. "Biorefineries as the base for accomplishing the sustainable development goals (SDGs) and the transition to bioeconomy: Technical aspects, challenges and perspectives." *Bioresour Technol* 340:125626. doi: 10.1016/j.biortech.2021.125626.

Solarte-Toro, Juan Camilo, and Carlos Ariel Cardona Alzate. 2021b. "Biorefineries as the base for accomplishing the sustainable development goals (SDGs) and the transition to bioeconomy: Technical aspects, challenges and perspectives." *Bioresource Technology* 340:125626. doi: https://doi.org/10.1016/j.biortech.2021.125626.

Solarte-Toro, Juan Camilo, Yessica Chacón-Pérez, and Carlos Ariel Cardona-Alzate. 2018. "Evaluation of biogas and syngas as energy vectors for heat and power generation using lignocellulosic biomass as raw material." *Electronic Journal of Biotechnology* 33:52-62. doi: https://doi.org/10.1016/j.ejbt.2018.03.005.

Steven, Soen, Elvi Restiawaty, and Yazid Bindar. 2021. "Routes for energy and bio-silica production from rice husk: A comprehensive review and emerging prospect." *Renewable and Sustainable Energy Reviews* 149:111329. doi: https://doi.org/10.1016/j.rser.2021.111329.

Su, Jialei, Feng Shen, Mo Qiu, and Xinhua Qi. 2017. "High-Yield Production of Levulinic Acid from Pretreated Cow Dung in Dilute Acid Aqueous Solution." *Molecules* 22 (2):285.

Suganya, T., M. Varman, H. H. Masjuki, and S. Renganathan. 2016. "Macroalgae and microalgae as a potential source for commercial applications along with biofuels production: A biorefinery approach." *Renewable and Sustainable Energy Reviews* 55:909-941. doi: https://doi.org/10.1016/j.rser.2015.11.026.

Susana Alcaraz Argumánez. 2013. "*Estudio tecno-energético de la producción de ácido levulínico a partir de celulosa mediante un proceso combinado: Descomposición hidrotérmica y deshidratación con catalizadores heterogéneos.*" [*Techno-energetic study of the production of levulinic acid from cellulose through a combined process: hydrothermal decomposition and dehydration with heterogeneous catalysts*]. Bachelor en Ingeniería de la Energía, Ingeniería de la Energía, Universidad Rey Juan Carlos. [Bachelor of Energy Engineering, Energy Engineering, Rey Juan Carlos University].

Takkellapati, S., T. Li, and M. A. Gonzalez. 2018. "An Overview of Biorefinery Derived Platform Chemicals from a Cellulose and Hemicellulose Biorefinery." *Clean Technol Environ Policy* 20 (7):1615-1630. doi: 10.1007/s10098-018-1568-5.

Terzopoulou, Zoi, Lazaros Papadopoulos, Alexandra Zamboulis, Dimitrios G. Papageorgiou, George Z. Papageorgiou, and Dimitrios N. Bikiaris. 2020. "Tuning the properties of furandicarboxylic acid-based polyesters with copolymerization: A review." *Polymers* 12. doi: 10.3390/POLYM12061209.

Thanigaivel, S., A. K. Priya, Kingshuk Dutta, Saravanan Rajendran, Karthikeyan Sekar, A. A. Jalil, and Matias Soto-Moscoso. 2022. "Role of nanotechnology for the conversion of lignocellulosic biomass into biopotent energy: A biorefinery approach for waste to value-added products." *Fuel* 322:124236. doi: https://doi.org/10.1016/j.fuel.2022.124236.

Thanigaivel, Sundaram, Sundaram Vickram, Nibedita Dey, Govindarajan Gulothungan, Ramasamy Subbaiya, Muthusamy Govarthanan, Natchimuthu Karmegam, and Woong Kim. 2022. "The urge of algal biomass-based fuels for environmental sustainability against a steady tide of biofuel conflict analysis: Is third-generation algal biorefinery a boon?" *Fuel* 317:123494. doi: 10.1016/j.fuel.2022.123494.

Tolosa, R. A., G. Jimenez-Obando, N. P. Arias, C. A. Cardona, and O. Giraldo. 2014. "Cementicious Materials Reinforcement Using Angustifolia kunth Bamboo Fiber Covered with Nanostructured Manganese Oxide." *Industrial & Engineering Chemistry Research* 53 (20):8452-8463. doi: 10.1021/ie403958y.

Ubando, Aristotle T., Charles B. Felix, and Wei-Hsin Chen. 2020. "Biorefineries in circular bioeconomy: A comprehensive review." *Bioresource Technology* 299:122585. doi: 10.1016/j.biortech.2019.122585.

Union, European. 2012. *Innovating for Sustainable Growth: A Bioeconomy for Europe.*

Vasiliev, Valery V., and Evgeny V. Morozov. 2018. "Introduction." In *Advanced Mechanics of Composite Materials and Structures (Fourth Edition)*, edited by Valery V. Vasiliev and Evgeny V. Morozov, xvii-xxv. Elsevier.

Verge, Pierre, Valérie Toniazzo, David Ruch, and João A.S. Bomfim. 2014. "Unconventional plasticization threshold for a biobased bisphenol-A epoxy substitution candidate displaying improved adhesion and water-resistance." *Industrial Crops and Products* 55:180-186. doi: 10.1016/j.indcrop.2014.01.048.

Vivien, F. D., M. Nieddu, N. Befort, R. Debref, and M. Giampietro. 2019a. "The Hijacking of the Bioeconomy." *Ecological Economics* 159 (June 2018):189-197. doi: 10.1016/j.ecolecon.2019.01.027.

Vivien, F. D., M. Nieddu, N. Befort, R. Debref, and M. Giampietro. 2019b. "The Hijacking of the Bioeconomy." *Ecological Economics* 159:189-197. doi: 10.1016/ j.ecolecon.2019.01.027.

Vu, Hang P., Luong N. Nguyen, Minh T. Vu, Md Abu Hasan Johir, Robert McLaughlan, and Long D. Nghiem. 2020. "A comprehensive review on the framework to valorise lignocellulosic biomass as biorefinery feedstocks." *Science of The Total Environment* 743:140630. doi: 10.1016/j.scitotenv.2020.140630.

Vy Tran, A., S.-K. Park, H. Jin Lee, T. Yong Kim, Y. Kim, Y.-W. Suh, K.-Y. Lee, Y. Jin Kim, and J. Baek. 2022. "Efficient Production of Adipic Acid by a Two-Step Catalytic Reaction of Biomass-Derived 2,5-Furandicarboxylic Acid." *Chem Sus Chem* 15. doi: 10.1002/cssc.202200375.

Wang, Shuguo., Zehui. Zhang, and Bing. Liu. 2015. Catalytic conversion of fructose and 5-hydroxymethylfurfural into 2,5-furandicarboxylic acid over a recyclable Fe_3O_4-CoO_x magnetite nanocatalyst. In *ACS Sustainable Chemistry and Engineering*.

Weber, Caroline Trevisan, Luciane Ferreira Trierweiler, and Jorge Otávio Trierweiler. 2020. "Food waste biorefinery advocating circular economy: Bioethanol and distilled beverage from sweet potato." *Journal of Cleaner Production* 268:121788. doi: 10.1016/j.jclepro.2020.121788.

Werpy, T., and G. Petersen. 2004. *Top Value Added Chemicals from biomass. No.: DOE/GO-102004-1992*. www.osti.gov/bridge: U.S. Department of Energy, Office of scientific and technical information.

Xie, Hongzhou, Linbo Wu, Bo-Geng Li, and Philippe Dubois. 2019. Modification of Poly(ethylene 2,5-furandicarboxylate) with Biobased 1,5-Pentanediol: Significantly Toughened Copolyesters Retaining High Tensile Strength and O_2 Barrier Property. In *Biomacromolecules*.

Yang ST., Zhang K., Zhang B., and Huang H. 2011. "Biobase chemicals: Fumaric acid." In *Comprehensive Biotechnology*, edited by Moo-Young and Bulter Michael, 163-77. The Netherlands.

Yao, Zhen, Pingping Zhou, Bingmei Su, Sisi Su, Lidan Ye, and Hongwei Yu. 2018. "Enhanced Isoprene Production by Reconstruction of Metabolic Balance between Strengthened Precursor Supply and Improved Isoprene Synthase in Saccharomyces cerevisiae." *ACS Synthetic Biology* 7:2308-2316. doi: 10.1021/acssynbio.8b00289.

Yong, Khai Jie, Ta Yeong Wu, Cornelius Basil Tien Loong Lee, Zhi Jin Lee, Qinpu Liu, Jamaliah Md Jahim, Qiaoqiao Zhou, and Lian Zhang. 2022. "Furfural production from biomass residues: Current technologies, challenges and future prospects." *Biomass and Bioenergy* 161:106458. doi: 10.1016/j.biombioe.2022.106458.

Zhang, Heng, Pau Cabañeros Lopez, Claire Holland, Alan Lunde, Morten Ambye-Jensen, Claus Felby, and Sune Tjalfe Thomsen. 2018. "The multi-feedstock biorefinery – Assessing the compatibility of alternative feedstocks in a 2G wheat straw biorefinery process." *GCB Bioenergy* 10 (12):946-959. doi: https://doi.org/10.1111/gcbb.12557.

Zhang, Ming, Li Xie, Zhixuan Yin, Samir Kumar Khanal, and Qi Zhou. 2016. "Biorefinery approach for cassava-based industrial wastes: Current status and opportunities." *Bioresource Technology* 215:50-62. doi: https://doi.org/10.1016/j.biortech.2016.04.026.

Zhang, Xin, Kaili Nie, Yongli Zheng, Fang Wang, Li Deng, and Tianwei Tan. 2015. "Enzymatic production and functional characterization of d-sorbitol monoesters with various fatty acids." *Catalysis Communications* 72:138-141. doi: https://doi.org/10.1016/j.catcom.2015.09.010.

Zou, Caineng, Qun Zhao, Guosheng Zhang, and Bo Xiong. 2016. "Energy revolution: From a fossil energy era to a new energy era." *Natural Gas Industry B* 3:1-11. doi: 10.1016/j.ngib.2016.02.001.

Chapter 3

Process Scale-Up for Biorefineries in Latin America: Advances and Challenges

Henry F. González[1], Melvin Durán-Rincón[1] and Valentina Aristizábal-Marulanda[1,2]

[1]Facultad de Tecnología, Escuela de Tecnología Química,
Grupo de investigación en Desarrollo de Procesos Químicos,
Universidad Tecnológica de Pereira, Pereira, Colombia
[2]Grupo de investigación en Procesos Agroindustriales y Desarrollo Sostenible (PADES),
Universidad de Sucre, Sincelejo, Colombia

Abstract

In a Latin American approach (Caribbean, Central and South America), Brazil is a focus of research and development of biorefining processes. Researchers from that country have done remarkable work on first and second generation of biofuels just like Colombia on first generation biofuels. The research has been associated to sugarcane and oil palm transformation and their main residues (e.g., vinasse, bagasse, empty fruit bunches, etc.). Therefore, dealing with lignin and cellulose materials remains the main research challenge. Biomass availability is related to agricultural production. The largest agricultural economies in Latin America after Brazil are Mexico, Argentina, and Colombia. Different biorefinery options have been proposed in these countries to treat agricultural waste. It applies both to extensive crops and to smaller-scale crops and their residues. In this sense, this work addresses some challenges for process scaling up, and the search for sustainable biorefineries in the Latin America context concentrating on the following topics, i) stabilization of biorefinery feed, ii) technoeconomic analysis focused on finding alternatives or technologies for viable projects and iii) analysis of stakeholders in biorefineries. Other important elements

In: The Future of Biorefineries
Editor: Waldemar Nyström
ISBN: 979-8-88697-524-6

discussed are the emerging microorganisms that could play a pivotal role in scaling process and the need to integrate environmental concepts and supply chain in biorefineries. In conclusion, there are opportunities for the development of biorefineries in Latin America taking advantage of the Brazilian experience and based on agricultural practices that may be common in the region. Research and development of new unit "bio" - operations should be strengthened.

Keywords: bioprocesses, residual biomass, biorefinery, Latin America context, scaling up

1. Introduction

Several efforts have been made in recent decades to reduce carbon emissions and other greenhouse gases. Biomass has been proposed as an alternative renewable carbon source to balance biomass fixation with carbon emission, both for energy transformation and to produce materials. (DeAngelo et al. 2021) have found that renewable sources would account for 60% of primary energy and at least half of that renewable energy would be provided by biomass.

Biorefineries have been proposed as integrated processes for the use of biomass, however, biorefineries continue to compete economically with oil derivatives, which have lower production costs and mature technologies in current commercial production. (Rosales-Calderon and Arantes 2019).

Biomass production for biorefineries has certain restrictions, for example, there is often pressured to comply with a sustainability framework; they must not threaten the food security of the population, compete for the use of land and other resources, or cause greater pollution problems than they are intended to prevent.

In addition to the pressure to comply with environmental regulations and the challenges of developing an immature technology, it must also compete with technologies that were not developed within sustainable production frameworks or a circular economy. So, the challenge for technological development is high.

Despite the enormous availability of biomass on the planet and its potential for energy production, the biomass of interest for biorefineries is based on residual biomass or waste.

Here we present a review of the research on biorefinery scale-up, that is, the technological maturation processes necessary to carry out biorefineries from their conception or from laboratory scale to commercial grade.

The largest proportion of biorefinery proposals in Latin America have the characteristics mentioned in Figure 1. Several key aspects for research in biorefinery scale-up are reviewed below.

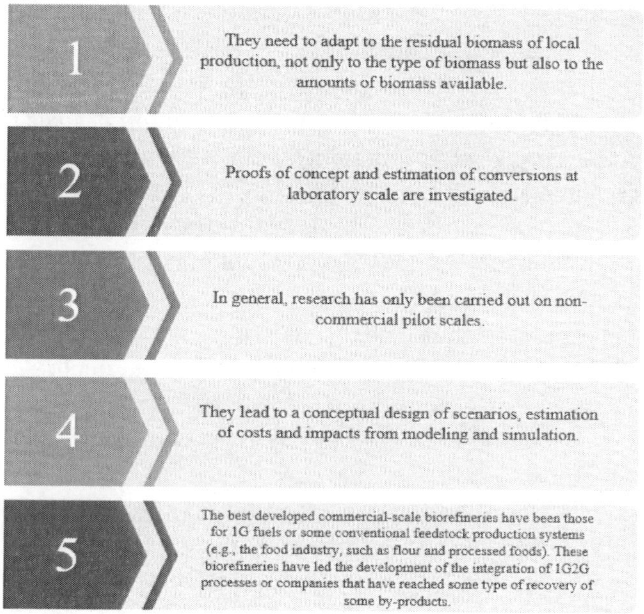

Figure 1. Characteristics of most biorefinery proposals in Latin America.

2. Biomass Availability and Potential Raw Materials

The development of biorefineries has been conditioned to the use of large amounts of agro-industrial waste, greater agro-industrial activity, greater amounts of waste, or materials derived from conventional production chains.

The availability of biomass is one of the crucial factors to feed a biorefinery, however, it does not necessarily imply that large-scale commercial systems should be designed. In fact, Cutz et al. (2016) considered the need for small-scale commercial technologies in Central America for the conversion of biomass to energy, in such a way that the raw material is available within a radius of 50 km, it means taking advantage of small-scale agricultural systems. However, small-scale systems decrease energy efficiency. Fuess and Zaiat (2018) also propose the design of compact plants to reduce the risks derived from the economic uncertainties of technological development.

Despite the need to design alternatives for smaller-scale economic activities, as in the case of orange peel, Ortiz-Sanchez et al. (2021) found that processing 280 kg/h of waste did not result in profitable options due to high manufacturing costs. This leads to the proposal of other alternatives such as the production of pellets for combustion.

Serna-Loaiza et al. (2018) required sizing a biorefinery for 50 t/day of cocoyam, however, the economic analysis identified that the smallest biorefinery that produced profitability was 880 t/day, which demanded increase agricultural activities.

Other alternatives that may arise, are biorefineries that take advantage of various types of waste from various agricultural activities, such as multi-feedstock or multi-substrate.

However, for each substrate the possible compounds that can act as inhibitors during bioprocesses must be analyzed, for example (Ortiz-Sanchez et al. 2021) states that it is useful to extract oils such as D-limonene from orange peel to prevent this compound interferes with microorganisms during anaerobic digestion.

When small-scale biorefineries are proposed in rural areas, Solarte-Toro et al. (2022) suggest that in addition to the design of the process, the environmental and social factors of the projects should be considered. Through methodologies such as LCA and social impact assessment (SIA)

Most of the derivatives of the agricultural industry are lignocellulosic biomass residues or materials that require complex processes for their use. The actors in the value chain of agricultural production ensure their income and profitability from a main or primary product, which can be agricultural products or agro-industrial products. At a next level, integrated processes are formulated for the use of derivatives, by-products, or residues.

The first biorefineries were linked to the production of first-generation fuels, which resulted from additional processing to the conventional processes of sugar and fat production in the food industry. However, the competition between food production and first-generation biofuels generated pressures on the costs of food, agricultural inputs, farmland demand, among other externalities. This has led to the formulation of fuel production alternatives from waste from agricultural production processes or second-generation biofuels.

In the case of first-generation ethanol, the sugars were simply extracted from the sugarcane after minor physical and chemical treatments to remove contaminants. The extracted sugars were concentrated and then fermented. In the case of second-generation alcohol production, the cellulose-hemicellulose-lignin matrix must first be broken, and then the cellulose and hemicellulose must be broken into pentoses and hexoses, using appropriate enzymes or catalytic processes (Chandel et al. 2019).

The main agro-industrial residues that are currently candidates for biorefineries are seeds, shells, milling by-products, molasses or residual syrups or spent products such as stillage. In addition to other production residues such as stems, fodder, straw, etc.

Most agricultural residues are a mixture of polysaccharides, polypeptides, fats, and other substances of biological origin that act as primary or secondary metabolites, such as phenolic compounds, carotenoids, etc. As an example, the most abundant waste from the agricultural industry, such as sugar cane bagasse, sugar cane straw, corn cob and rice straw contain 53.5%, 58.7%, 58.6% and 61.04% of total carbohydrates (cellulose and hemicellulose) and 27.60%, 31.10%, 18.80% and 19.86% of lignin, respectively (Chandel et al. 2019)

The possibility of obtaining products from lignocellulosic materials also raises the possibility of starting from wood, such is the case of the approach of processes to produce hydrogen by gasification and dark fermentation using *Pinus patula* (García-Velásquez et al. (2017). Several factors must be considered for the biomass availability as shows Figure 2.

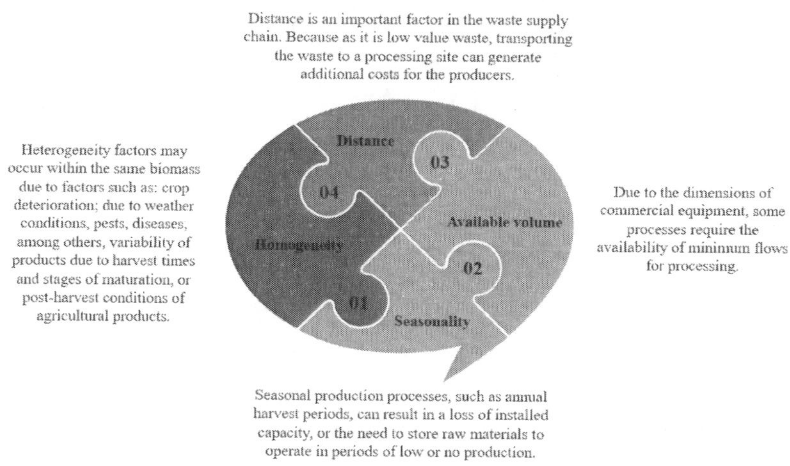

Figure 2. Relevant factors that affect the biomass availability.

3. Stabilization of the Input Feed

Due to the conditions described above, the operation of a biorefinery requires stable conditions. To avoid disturbances in the process, it is necessary to stabilize and regularize the composition of the raw material of the biorefineries. These situations are less obvious in small-scale tests, due to the sample size required. But to the extent that the processes are scaled, the problems of heterogeneity of the substrates will appear.

Biorefineries contain transformation stages based on enzymes or fermentations, therefore, the composition of the medium (pH, nutrients, chemical species, presence of microorganisms, etc.) are determining factors in the efficiency of biological reactions. Therefore, the stabilization of the substrate will be vital for the functioning of the bioprocesses.

The pretreatments were developed as an alternative to increase the solubilization of polysaccharides or for the fractionation of lignocellulosic material.

The pretreatment stages are bottlenecks for the development of biorefineries because they generate costs related to the use of chemical products, energy to pressurize or heat, in addition, most of the pretreatments are potential generators of inhibitory byproducts, such is the case of subcritical water hydrolysis of rice straw, studied by (Abaide et al. 2019) in which furfural and HMF appear above 180 C and a pressure of 25 MPa.

In the last decade several laboratory-scale pretreatment processes have been reported, however, the scaling of pretreatment processes to pilot plant sizes has been reported by Brenelli et al. (2022) for steam explosion of sugarcane straw at 200°C, 15 bar (between 1 and 10 min) for conversion to xylooligosaccharides.

For this reason, research in recent years has focused on proposing smoother and more efficient pretreatment processes, seeking to maximize sugar recovery, reduce energy costs, the use of chemical products and avoid unwanted by-products.

Enzymatic treatments are the natural alternative (or based on biomimetics) for the release of sugars, enzymatic reactions can occur at room temperature, however, the use of enzymes from extremophilic organisms can be used at temperatures greater than 50°C, which is favorable for carrying out reactions at an industrial level, by inhibiting the growth of unwanted microorganisms, without using strict sterilization protocols.

To date, the development of research has provided a better understanding of enzymatic processes, the activity of enzymes and their role in the depolymerization of polysaccharides. Also, to counteract the high cost of enzymes, progress has been made in the production of crude enzymes in situ. Despite the above, it is still necessary to use physical, mechanical, chemical pretreatments or a mixture of them to optimize enzymatic hydrolysis.

Ellilä et al. (2017) As proposed alternatives to produce cellulases with low cost and optimal performance, the fungus *Trichoderma reesei* M44 was initially tested, cultivated on various waste substrates from the Brazilian agricultural industry, such as soybean hulls, bagasse, and cane molasses. Productions higher than 20 g/L of extracellular enzyme were obtained. Because the presence of sugars in molasses (sucrose, fructose, and glucose) represses cellulase production in microorganisms, alternatives such as genetic transformation are sought. Enzymes are induced in the presence of reducing sugars, and additionally, the expression of heterologous beta-glucosidase enzymes from the fungus *Talaromyces emersonii*.

In the same study, the enzymes are produced by the fungus grown on soybean hulls and sugar cane molasses and were tested for the saccharification of sugar cane bagasse, and a conversion of glucose and xylose of almost double than obtained by the fungus native strain. Transformation processes were also used for a regulated expression of *Aspergillus niger* invertase by the transforming strain of *T. ressei,* avoiding the use of acid to hydrolyze sucrose from molasses. This tests the potential use of genetic tools to optimize enzyme production and saccharification processes in biorefining processes.

4. Emerging Microorganisms

Research with microorganisms is key in biorefinery processes and is based on the following interests:

- Algae.
- Microorganisms producing lignocellulosic enzymes.
- Fermentative microorganisms adapted to the substrates.
- Microorganisms capable of synthesizing products of interest.

Algae are an important field of interest in Latin America, the ability of these microorganisms to generate biomass can enhance the feeding of biorefineries. The possibility of fixing CO_2 represents an environmental potential, the metabolic capacities of the algae allow to take advantage of the biomass of many residues. However, factors such as operating costs due to energy requirements for cultivation, harvesting, cell lysis, fuel synthesis and inputs for algae cultivation can become restrictions for the implementation of technologies (Klein et al. 2019).

Klein et al. (2019) proposes the use of algae in a biorefinery concept integrated with the cultivation and processing of sugar cane to produce fuels with reduced CO_2 emissions. However, in a study of algae cultivation using pre-treated domestic wastewater as a substrate, Assis et al. (2019) found that the supply of industrial exhaust gases (rich in CO_2) represents high investment costs in the algae cultivation process. In addition, the demand for CO_2 in the crop is low due to the high concentration of carbon in the substrate.

The use of vinasse for the growth of microalgae was studied by (Santana et al. 2017). Vinasse is a residue of ethanol production, it is estimated that for each liter of ethanol, between 12 and 14L of vinasse are produced. The authors searched for highly productive microalgae strains capable of growing in sugarcane vinasse and meeting the requirements for algal growth and nutrient uptake. Two strains of interest were found within a collection of 40 strains. The selected algae were *Micractinio sp.* and *Chlamydomonas biconvexa*. These microorganisms in a 12-h light/dark regimen can use the glycerol carbon for growth in axenic culture. In a non-auxenic culture, other carbon sources such as carbohydrates can be metabolized. With this process, important amounts of biomass rich in proteins and carbohydrates can be obtained.

Efforts to produce enzymes has carry out in a 5 L Stirred-Tank Bioreactor (STR) to produce laccases, peroxidases, and manganese peroxidases by *Marasmiellus palmivorus*, focusing on enzymatic hydrolysis second generation or ethanol production (Schneider et al. 2018)

Production of pectinase has been tested by culturing *Aspergillus niger* in a solid-state fermentation in a 200 L (bed capacity) pilot scale packed bed bioreactor, pectinase is used in the saccharification of pectin from citrus waste. Finkler et al. (2017) found that agitation to avoid agglomeration is necessary to uniform the levels of pectinase produced.

The development and simulation of bioprocesses to produce poly(3-hydroxybutyrate) from dairy wastewater using *Bacillus subtilis* EPAH18 was analyzed (Peña-Jurado et al. 2019), where the data was tested in a benchtop STR (3 L). The cost of production is calculated and estimated to compare it with other processes, and the flow of cheese whey necessary to feed the process is defined. The authors found opportunities for the process but conclude that more efforts should be made to scale up the process.

5. Techno-Economic Analysis

Although the first alternative to treat the waste results in final disposal, environmental regulations have pressured producers to look for alternatives such as composting and, in a more sophisticated way, direct combustion for the generation of heat or electricity, whenever meet the following conditions:

- CAPEX investments can be leveraged on the existence of a previous economic activity based on a conventional production,
- There is an adequate return on investment, and this depends on cost competition for grid power generation and the costs of conventional fuels.
- The availability of mature technologies for obtaining energy from waste and, consequently, the reduction of investment risk.

The techno-economic analysis is a fundamental stage for the proposal of new processes, the methodologies used by different authors for the economic evaluation are very similar and are based on ratios for evaluation of projects such as the Net Present Value (NPV), the Internal Rate of Return (IRR), or the recovery period. These indicators are based on the estimation of the costs of the processes, the projection of the sales prices of the products obtained and the profit margins (Watanabe et al. 2016).

Cost estimation is a good approximation for the theoretical scaling of processes, however, it contains uncertainties, especially when the processes are in early stages of technological maturity, because most of the information arises from performance obtained in experimental-scale processes, or other similar processes on the market. Even so, the techno-economic projections are an important tool to analyze the process, they allow to identify restrictive costs and with this, to propose alternatives.

As an example, the case of Fuess et al. (2017) who proposed the Anaerobic Digestion of Vinasse optimized by alkalinization, later and thanks to the design and cost evaluation tools, the same research group (Fuess et al. 2018) propose an Anaerobic Digestion in two phases, with a phase acidogenic that produces a gas rich in Hydrogen, in a process with less demand for sodium hydroxide for alkalization, and improving the electrical energy produced between 20% and 30%.

The development of several biorefineries to date has been built around conventional economic activities such as food production and this includes the establishment of first-generation biofuel biorefineries. This means that agribusiness companies can take advantage of potential investments in biorefineries.

Investment in biorefineries that obtain waste from suppliers, or third parties would have the negative aspect of losing bargaining power vis-à-vis waste generators (Aboytes-Ojeda et al. 2022), for example, setting the price of waste could affect the profitability of the business.

Complementary to the economic analysis, sensitivity analyzes of economic factors have been proposed (Fuess and Zaiat 2018) and more recently risk measures such as the Conditional Value at Risk (CVaR) (Aboytes-Ojeda et al. 2022) applied to the optimization of the supply chain, which takes as input the total capital investment, production capacity and performance, and other costs such as transportation and factors such as demand.

Chandel et al. (2019) recognize that second generation biorefineries can complement primary production, taking advantage of strategic benefits such as the co-use of equipment, facilities, transportation, and other factors that can substantially reduce costs. However, they recognize that different activities have different business models. From a holistic perspective, process development should also involve value chain analysis and business model propositions.

However, companies that operate multiple business units also compare financially the rates of return on invested capital and have a natural preference to invest in the most profitable business units, if the most profitable activity is the main one; the development of biorefineries will encounter many investment barriers to their development. On the other hand, it is desirable that the biorefinery be the one that generates the highest profits, in this way the investment in the primary activity would be leveraged by the products of the biorefinery. Table 1 indicates examples of biorefineries in Latin America context, raw materials, processes, products, scale, and analysis.

5.1. Design *vs.* Economic Performance

The conceptualization of facility design alternatives as possible scenarios is evaluated and cost sensitivity analyzes are proposed to deal with uncertainties, for example the one proposed by (Fuess and Zaiat 2018) who determined specific factors for the case of anaerobic digestion. of the vinasse, identified that such sensitivity factors were investment costs, operating costs, alkalinization cost, electricity price, biomethane price and distance to the methane distribution network.

Table 1. Recent research papers in Biorefineries scaling up in Latin America

Substrate	Process	Products	Scale base	Analysis	Country	Reference
Sugarcane vinasse and glycerol	Co-digesting	Methane	Lab scale	Energy potential assessment	Brazil	(Borges et al. 2021)
Sugarcane bagasse (Raw and hydrothermally pretreated)	Biomass Washing, Alkaline Treatment, Enzymatic Hydrolysis, Alcoholic Fermentation	Ethanol	Lab scale	Yields. Simulation, Techno-Economic Analysis	Brazil	(Pinto et al. 2021)
Residual biomass of banana (harvest and commercialization)	Co-culture fermentation (*Saccharomyces cerevisiae* and *Pachysolen tannophilus*)	Fuel bioethanol	Lab scale	Yields, simulation, logistic-economic assessment	Brazil	(Uchôa et al. 2021)
Chlorella vulgaris biomass	Multistage liquid-liquid extraction, Dilute sulfuric acid hydrolysis	Lipids, proteins, pigments and carbohydrates	Lab scale	Yields	Brazil	(Martins et al. 2021)
Corncob	Olid-state fermentation (SSF) of *Phanerochaete chrysosporium*	Ligninolytic enzymes (Lignin peroxidase, Manganese Peroxidase, and Laccase)	Lab scale	Culture conditions optimization	Mexico	(Sosa-Martínez et al. 2021)
Sugarcane bagasse	Pretreatment, enzymatic hydrolysis, ABE fermentation, downstream process	Acetone, Butanol, Ethanol and, Xylitol, electricity	Lab scale	Simulation, Economic Analysis	Mexico	(Morales-Rodriguez et al. 2021)
Unbleached pine sawdust pulp		Microfibrillated cellulose	Lab scale	Characterization	Argentina	(González et al. 2022)
Agave lechuguilla Bagasse	Drying Procedure and Enzymatic Hydrolysis	Bioactive compounds (Flavonoids), biogas and biofuel	Lab scale	Yields	Mexico	(Morreeuw et al. 2021)

Table 1. (Continued)

Substrate	Process	Products	Scale base	Analysis	Country	Reference
Sugarcane bagasse biochar (Biofertilizer) and vinasse	Microalgae cultivation *Chlorella vulgaris* and *Desmodesmus* sp.	Biomass (lipids and carbohydrates)	Lab scale	Yields	Brazil	(Ferreira et al. 2021)
Waste from plantain agro–Industry (rejected unripe plantain fruits (*Musa* AAB Simmonds))	Liquefaction, saccharification, and transglycosylation (Enzymatic)	Isomalto–oligosaccharides (IMO)	Lab scale	Simulations, Techno–economic analysis (NPV)	Colombia	(Gómez, Berni, et al. 2022)
Pine sawdust	Soda-ethanol pretreatment, simultaneous saccharification and fermentation	Bioethanol	Lab scale	Yield and modelling	Argentina	(Mendieta et al. 2022)
Wheat straw	Hydrothermal pretreatment	Xylan and glucan	Lab scale	Yields	Mexico	(Rodríguez et al. 2022)
Sugarcane vinasse	Biphasic extraction with ethyl acetate and ad	Trans-aconitic acid, and biogas	Lab scale	Yields	Colombia	(Montoya et al. 2021)
Sugarcane bagasse hydrolysates and molasses	Fermentation by fixed-bed reactor with immobilized recombinant s. Cerevisiae	Ethanol	Lab scale 100 mL	Yields	Brazil	(Perez et al. 2022)
Ash from oil palm	Microalgae growth	Biomass from microalgae	Lab scale 3 L Erlenmeyer flasks and pilot scale 120L	Yields	Brazil	(Tessari et al. 2022)
Tequila vinasse waste	Yeast culture	Yeast-based protein	Lab scale 5 L benchtop bioreactor	Yields	Mexico	(Díaz-Vázquez et al. 2022)
Pine sawdust	Soda–ethanol delignification		Lab scale 7 L pressurized reactor	Yields	Argentina	(Imlauer Vedoya et al. 2022)

Substrate	Process	Products	Scale base	Analysis	Country	Reference
Pine sawdust	Soda–ethanol delignification		Lab scale 7 L pressurized reactor	Yields	Argentina	(Imlauer Vedoya et al. 2022)
Vinasse	Two-stage anaerobic digestion	Hydrogen and methane	Lab scale. Anaerobic fluidized bed reactor total volume of 1.980 L	Yields (energy)	Brazil	(Ramos et al. 2022)
Vinasse, filter cake and deacetylation liquor (waste stream from the pretreatment of sugarcane straw for the 2G ethanol production)	AD	CH4	Lab scale. Batch reactor of 0.5 L capacity	Yields	Brazil	(Volpi et al. 2022)
Sugarcane vinasse, filter cake, and deacetylation liquor	Co-digestion	Ch4	Lab scale. Semi-continuous stirred tank reactor (s-cstr) consisted of a 5 L-duran flask with 4 L-working	Microbial community characterization, reactor efficiency, yields	Brazil	(Volpi et al. 2021)
Algae *Macrocystis pyrifera*, *Lessonia vadosa* and Corn Starch		Sorbitol, Isosorbide, Energy, Heat Power, H2, Fertilizers	Laboratory, 50 mL reactor	Mixed-integer nonlinear programming (MINLP), The Renewable Process Synthesis Index Metric (repsim), NVP	Argentina	(Pedrozo et al. 2022)
Brewer's spent grains	Subcritical water hydrolysis	Sugars	Lab-scale (test)	Scaling up analysis, economic assessment, project feasibility, sensitivity analysis, gantt chart	Brazil	(Sganzerla et al. 2022)

Table 1. (Continued)

Substrate	Process	Products	Scale base	Analysis	Country	Reference
Sugar cane	Gasification catalytic conversion (Fischer-Tropsch reactions)	Renowable jet fuel	Modeling	Simulation, net present value (NPV), the internal rate of return (IRR), as well as the biofuel production costs, The Life Cycle Assessment (LCA)	Brazil	(Guimarães et al. 2022)
Anaerobic sewage sludge	Pyrolysis	Syngas, bio-oil, biochar	Modelling	Kinetic and thermodynamic models	Brazil	(Alves et al. 2022)
Avocado (*Persea Americana* Mill)	Process conceptualization	Guacamole, avocado oil, animal feed, and electricity	Modelling	Simulation, The economic assessment (capex, opex, Payback period, NPV, and Minimum Processing Scale for Economic Feasibility MPSEF)	Colombia	(Solarte-Toro et al. 2021)
Orange peel waste	Biorefinery conceptualization	Essential oil, hesperidin, pectin, acetone, butanol, ethanol, biogas, and fertilizer.	Modelling	Simulation, lca, sensitivity analysis	Colombia	(Ortiz-Sánchez et al. 2021)
Seaweed *Sargassum muticum*	Biorefinery conceptualization	Organic fertilizer	Modelling	Economic analysis	Brazil	(Caxiano et al. 2022)
Conventional sugarcane (CS) and its straw recovered from the field, energy-cane (EC), and eucalyptus (Eu)	Steam Rankine cycle, integrated gasification combined cycle (IGCC)	Electricity	Modelling	Economic Assessment (project viability using discounted cash flow), Environmental assessment using LCA	Brazil	(Bressanin, Guimarães, et al. 2021)

Substrate	Process	Products	Scale base	Analysis	Country	Reference
Eucalyptus grandis	Biorefinery conceptualization	Ethanol	Modelling	Energy return on investment (EROI), carbon footprint (CF)	Uruguay	(Rachid-Casnati et al. 2021)
Shrimp	Biorefinery conceptualization	Chitin, chitosan and astaxanthin.	Modelling	Techno-economic analysis, environmental assessment (traci 2.1 tool)	Colombia	(Zuorro et al. 2021)
Sugarcane and eucalypt	Fischer-Tropsch (FT) liquid biofuels production	S green diesel (GD), green gasoline (GG), and renewable jet fuel (RJF)	Modelling	Climate change and agricultural land occupation impacts, Economic assessment, Optimization	Brazil	(Bressanin, Geraldo, et al. 2021)
Sugarcane	Biorefinery conceptualization	Ethylene glycol, energy, heat, and ethanol	Modelling	Simulation, technoeconomic assessment (OPEX, CAPEX, NVP, Cash flow, Sensitivity analysis)	Brazil	(Bastos et al. 2022)
Avocado peel and seeds	Supercritical Fluid Extraction (SFE)	Polyphenolic compounds, ethanol and chemical products	Modelling	Simulation and Estimation of Economic Indicators	Colombia	(Restrepo-Serna et al. 2022)
Avocado (*Persea americana* var. Americana)	Biorefinery conceptualization	Guacamole, animal feed, dry biogas, electricity	Modelling	Simulation LCA, social impact assessment (SIA)	Colombia	(Solarte-Toro et al. 2022)
Agricultural waste	Multi-objective stochastic mixed integer nonlinear programming model	Biofuel	Modelling	Supply chain network design (scnd)	Colombia	(Morales-Chávez 2021)
Sugarcane bagasse	Hydrothermal liquefaction		Modelling	Techno-economic assessment and sensitivity analysis	Brazil	(Deuber et al. 2021)
Sugar cane	Biorefinery conceptualization	Electricity and chemicals	Modelling	Minimize the total cost of a network of processing, transportation, and storage instances by mixed-integer linear model (milp)	Argentina	(Machin Ferrero and Mele 2022)

Table 1. (Continued)

Substrate	Process	Products	Scale base	Analysis	Country	Reference
Waste of Dominico–Hartón plantain	Biorefinery conceptualization	Nutrient blocks and silage	Modelling	Simulations, Techno–economic analysis (NPV)	Colombia	(Gómez, Nobre, et al. 2022)
Sugarcane bagasse	Yeast culture in multilayer packed-bed bioreactor	Nanocellulose	Pilot plant	Yields	Brazil	(Katayama et al. 2022)
Pig manure (PM), coffee mucilage (CFM), and cocoa mucilage (CCM).	Dark Fermentation	Biohydrogen and Volatile Fatty Acids	Pilot plant (4 L reactor)	Yields and composition	Colombia	(Rangel et al. 2022)
Organic fraction of municipal solid waste (OFMSW), cooking oil		Energy, biofuels, organic fertilizers, and value-added chemical compounds (Glycerin, 1,3 propanediol	Pilot plant 40 L batch for biodiesel production and 9.6 m^3 for biodigester	Yields, quality and quantity, preliminary economic analysis	Brazil	(de Sousa et al. 2022)
Sugarcane bagasse	Steam explosion and xylooligosaccharide (xos) extraction, alkaline treatments and lignin precipitation, fermentation	Prebiotic xylooligosaccharides, cellulosic ethanol, cellulose nanofibrils and lignin nanoparticles	Pilot scale (continuous steam explosion unit, bagasse at a rate of 11.7 kg/h)	Yields	Brazil	(Pereira et al. 2021)
Swine wastewaters	Microalgae harvesting	Amino acids, fatty acids, and peptides	Pilot scale 500 L reactors	Yields and quality	Brazil	(Michelon et al. 2022)
Sugarcane straw	Steam explosion pretreatment	Xylo-oligosaccharides (prebiotics), fermentable sugars, bioethanol, and bioenergy production	Pilot-scale reactor 30 L	Yields	Brazil	(Brenelli et al. 2022)

However, the assumptions in terms of costs for processes in the early stages of technological development at the process design scale do not consider scaling up aspects, which must be progressively developed at the technological maturity pathway, due to many aspects that are not considered, such as:

- Mass transfer issues.
- Design of equipment and piping materials.
- Learning or experience curves, considering that many processes are not common in the market, the development of knowledge is required until full steady production is achieved. For example, treating scaling phenomena in biorefineries (de Mello, Sperança, and Pereira 2022).
- Investments in technological development are generally not considered.

Beyond the financial criterion, exploring the interests of the stakeholders, it is important that the research identifies and responds to the criteria of the decision makers. (Cutz et al. 2016) used a Fuzzy Multi-Actor Multi-Criteria Decision-Making (MCDM) methodology to determine the key aspects to consider when designing an energy system based on biomass in Central America. Finding that the most important aspects for decision makers for technology evaluation were the availability of resources, the competence of personnel, equipment, engineering companies and incentives/subsidies. Therefore, it is also necessary to consider other criteria for the design, beyond economic profitability.

5.2. Supply Chain

The availability of the raw material, in volume and seasonality, greatly determines the design of biorefineries. The development of biorefineries in extensive planting areas where it is possible to collect the raw material without incurring high transportation costs (or large storage silos) has been notable.

Supply chain optimization analyzes were performed by (López-Díaz et al. 2018) to identify potential sites for the establishment of biorefineries, considering factors such as water resources and agricultural activity in a region. This optimization considers objective functions of economic benefits, water requirements and CO_2 emissions.

6. Valorization and Downstream Process

The development of biorefineries has revolved around the production of energy, either by direct processes of heat or electricity production, through the supply of biogas or liquid biofuels. Biorefineries must face the challenge of:

- Produce high value products.
- Products that compete in production costs against conventional substrates, for example, petroleum derivatives, molasses. Not only in the cost of raw material, but in processing costs and yields.

On the other hand, the need to obtain higher value products leads to the formulation of downstream refinement and purification processes, which:

- They may require different needs than other products obtained by conventional chemical synthesis (Sánchez Muñoz et al. 2022),
- requires experience,
- They require a specialized technological offer.
- Significantly influence CAPEX and OPEX (Menon and Rao 2012).

Most of the proposed biorefineries have focused on primary production operations, but the separation and purification operations known as downstream require more research, there are not many publications (Branco-Vieira et al. 2020) (Costa et al. 2018) that cover these processes in detail in Latin America.

7. Environmental Issues

The designs for the scaling up of biorefineries include analyzes of the environmental impacts of different process scenarios. The analyzes allow obtaining alternatives that reduce these with reference to business-as-usual scenarios.

However, much of the information used for environmental analyzes is obtained from process modeling and baseline environmental impact information for each stage of the process. This reference information is recorded in databases that must be adjusted to the conditions of Latin America and even to the conditions of each one of the regions within a country.

Klein et al. (2018) proposed techno-economic and environmental assessments focused on specific solutions, such as replacing a fraction of jet fuel with alternative fuels with specifications approved by ASTM, which led to proposing and evaluating alternatives. They selected esters and hydroprocessed fatty acids as a viable alternative to replace 5% of Brazilian annual national aviation fuel consumption in 2014; This alternative is based on the processing of palm, macauba or soybean oils. Other scenarios also evaluated, such as the Fischer-Tropsch processes, or the gasification of lignocellulosic material from sugar cane or eucalyptus, alcohol jet (ATJ); so, it converts ethanol or isobutanol into jet fuel. The authors determine that, to meet the goal of ensuring carbon-neutral growth for international flights originating in Brazil, between 630 and 830 million liters per year of renewable jet fuel must be produced, which would represent an increase in production soybean farm from 33 million hectares to 55 million hectares.

(Fuess et al. 2018) investigated the anaerobic digestion of vinasse, finding a potential for electricity production, but evaluating the specific impacts of the use of vinasse digestate on the concentration of sodium in the soil when used as fertilization material. Since this process can concentrate the salts, even more so when sodium-based alkaline materials are used for alkalinization during anaerobic digestion. (Fuess et al. 2017) proposed that alkalinization of the Anaerobic Digestion process can optimize electricity production costs, either by using sodium hydroxide or by recirculating the effluent. That is an example of how to overcome possible environmental problems, through engineering controls.

Klein et al. (2019) consider the use of algae in biorefineries integrating with the production of biofuels from sugar cane, but they also consider the advantages of carbon bonds, implemented by the government program RenovaBio, within the economic analysis. In accordance with Grangeia, Santos, and Lazaro (2022) The RenovaBio program proposes the LCA methodology as a mechanism for estimating CO_2 and a framework for participating in the carbon market. Although for the moment, due to the post-pandemic, and the macroeconomic context, volatility and uncertainties in fuel prices remain. According to the authors, the CO_2 credit program in Brazil is still in a raw phase and in implementation.

One of the challenges to consider in the development of biorefineries is that the increase in the complexity of biorefineries or the increase in the added value of the products also increases the environmental impacts of the projects. (Ortiz-Sánchez et al. 2021).

8. Stakeholder Considerations

From an academic point of view, it is common to find biorefinery scaling initiatives, many of them based on laboratory-scale tests or scales smaller than pilot tests. Where the costs of large-scale designs are projected, as well as their environmental impacts. But it is also proposed to include other considerations within the formulation of the projects, for this it is necessary to characterize the social networks of the interested parties, the identification of their interests, the restrictions, and the expectations of each of the interested parties.

Several biorefinery proposals have used tools to integrate social variables in the projects, such is the case of Solarte-Toro et al. (2022) that have used Social Impact Assessment (SIA) techniques, considering factors such as employment generation and the availability of experienced personnel in rural areas, considering the design of small-scale biorefineries. Aristizábal-Marulanda et al. (2020) had already considered social actors such as workers and local communities and evaluated subcategories such as child labor, fair wages, minimum wages, working time, access to material resources, safe living conditions and healthy and local employment.

Chandel et al. (2019) highlight alliances at all levels or collaboration between the executing parties of the project, the supply chain, production, services and utilities, marketing, etc., as a factor of greater strategic importance. Even in the value chain to produce second-generation methanol from sugarcane bagasse, agents such as investors, corporate clients, co-development of processes with clients, suppliers and the creation of a business ecosystem are proposed.

Silva-Martínez et al. (2020) identified that, in the Latin American and Caribbean region, bio-waste (household organic waste and forest and agricultural waste) is not sufficiently recognized as a valuable energy source and that it is also an underutilized resource in the region. This is part of the Social Barriers for the execution and proposal of projects.

Conclusion

The development of biorefineries should be promoted, but it is necessary to turn the restrictions into challenges, such as:

- Make refineries profitable on a small scale. For any investor, it is better to have a low investment risk and gradually expand capacity.
- According to the above, the aspects that impede the economic viability of small-scale biorefineries must be identified.
- The development of biorefineries must be done in a modular way and from a central process of less complexity, the development of cash flows can leverage the expansion of refining, seeking higher value products.
- It is necessary to create a business ecosystem of suppliers and competitors around the supply of equipment, process units, engineering, etc.
- Leverage should be sought for research in process scale-up, a pilot plant can have several functions such as: process research, technology demonstration or small-scale production. However, a pilot plant can be a considerable investment with a very high margin of uncertainty for the investor.
- The development of biorefineries requires strategic investments in R&D leveraged with public funds but managed in programs with models such as the Innovative Partnership Program and Approving the work in phases.
- Based on the above, the Latin American countries that share production systems and agricultural products can propose a knowledge and technological development market to advance in the technological maturity of biorefineries, create clusters and improve competitiveness.
- At the supply chain level, intermediate crude products that can be processed in small production units and then centralized in biorefineries, such as hydrolysates, crude oil extracts, protein extracts, among others.

Acknowledgments

The authors express their gratitude to the project research called "Aprovechamiento y valorización sostenible de residuos sólidos orgánicos y su posible aplicación en borrefinerías y tecnologías de residuos a energía en el departamento de Sucre" with BPIN code 2020000100189, funded by the Sistema General de Regalías - SGR.

References

Abaide, Ederson R., Sérgio R. Mortari, Gustavo Ugalde, Alexsandra Valério, Suélen M. Amorim, Marco di Luccio, Regina de F. P. M. Moreira, et al. 2019. "Subcritical Water Hydrolysis of Rice Straw in a Semi-Continuous Mode." *Journal of Cleaner Production* 209 (February): 386–97. https://doi.org/10.1016/J.JCLEPRO.2018.10.259.

Aboytes-Ojeda, Mario, Krystel K. Castillo-Villar, and Yajaira Cardona-Valdés. 2022. "Bi-Objective Stochastic Model for the Design of Biofuel Supply Chains Incorporating Risk." *Expert Systems with Applications* 202 (September): 117285. https://doi.org/10.1016/J.ESWA.2022.117285.

Alves, J. L. F., J. C. G. da Silva, M. P. Languer, L. Batistella, M. di Domenico, V. F. da Silva Filho, R. F. P. M. Moreira, and H. J. José. 2022. "Assessing the Bioenergy Potential of High-Ash Anaerobic Sewage Sludge Using Pyrolysis Kinetics and Thermodynamics to Design a Sustainable Integrated Biorefinery." *Biomass Conversion and Biorefinery* 12 (3): 693–704. https://doi.org/10.1007/s13399-020-01023-2.

Aristizábal-Marulanda, Valentina, Juan Camilo Solarte-Toro, and Carlos Ariel Cardona Alzate. 2020. "Economic and Social Assessment of Biorefineries: The Case of Coffee Cut-Stems (CCS) in Colombia." *Bioresource Technology Reports* 9 (February): 100397. https://doi.org/10.1016/J.BITEB.2020.100397.

Assis, Tamires Condé de, Maria Lúcia Calijuri, Paula Peixoto Assemany, Alexia Saleme Aona de Paula Pereira, and Márcio Arêdes Martins. 2019. "Using Atmospheric Emissions as CO2 Source in the Cultivation of Microalgae: Productivity and Economic Viability." *Journal of Cleaner Production* 215 (April): 1160–69. https://doi.org/10.1016/J.JCLEPRO.2019.01.093.

Borges, A. D. V., L. T. Fuess, I. Alves, P. Y. Takeda, and M. H. R. Z. Damianovic. 2021. "Co-Digesting Sugarcane Vinasse and Distilled Glycerol to Enhance Bioenergy Generation in Biofuel-Producing Plants." *Energy Conversion and Management* 250. https://doi.org/10.1016/j.enconman.2021.114897.

Branco-Vieira, Monique, Sergio San Martin, Cristian Agurto, Marcos A. V. Freitas, António A. Martins, Teresa M. Mata, and Nídia S. Caetano. 2020. "Biotechnological Potential of Phaeodactylum Tricornutum for Biorefinery Processes." *Fuel* 268 (May): 117357. https://doi.org/10.1016/J.FUEL.2020.117357.

Brenelli, L. B., R. Bhatia, D. T. Djajadi, L. G. Thygesen, S. C. Rabelo, D. J. Leak, T. T. Franco, and J.A. Gallagher. 2022. "Xylo-Oligosaccharides, Fermentable Sugars, and Bioenergy Production from Sugarcane Straw Using Steam Explosion Pretreatment at Pilot-Scale." *Bioresource Technology* 357. https://doi.org/10.1016/j.biortech.2022.127093.

Brenelli, Lívia B., Rakesh Bhatia, Demi T. Djajadi, Lisbeth G. Thygesen, Sarita C. Rabelo, David J. Leak, Telma T. Franco, and Joe A. Gallagher. 2022. "Xylo-Oligosaccharides, Fermentable Sugars, and Bioenergy Production from Sugarcane Straw Using Steam Explosion Pretreatment at Pilot-Scale." *Bioresource Technology* 357 (August): 127093. https://doi.org/10.1016/J.BIORTECH.2022.127093.

Bressanin, J. M., V. C. Geraldo, F. D. A. M. Gomes, B. C. Klein, M. F. Chagas, M. D. B. Watanabe, A. Bonomi, E. R. D. Morais, and O. Cavalett. 2021. "Multiobjective Optimization of Economic and Environmental Performance of Fischer-Tropsch Biofuels Production Integrated to Sugarcane Biorefineries." *Industrial Crops and Products* 170. https://doi.org/10.1016/j.indcrop.2021.113810.

Bressanin, J. M., H. R. Guimarães, M. F. Chagas, I. L. D. M. Sampaio, B. C. Klein, M. D. B. Watanabe, A. Bonomi, E. R. D. Morais, and O. Cavalett. 2021. "Advanced Technologies for Electricity Production in the Sugarcane Value Chain Are a Strategic Option in a Carbon Reward Policy Context." *Energy Policy* 159. https://doi.org/10.1016/j.enpol.2021.112637.

Caxiano, I. N., P. A. Mello, P. H. R. Alijó, L. V. Teixeira, R. F. Cano, J. G. S. S. Maia, J. B. V. Bastos, and M. S. G. Pavão. 2022. "Continuous Design and Economic Analysis of a Sargassum Muticum Biorefinery Process." *Bioresource Technology* 343. https://doi.org/10.1016/j.biortech.2021.126152.

Chandel, Anuj K., Juliana Q. Albarelli, Diego T. Santos, Shishir Ps Chundawat, Munish Puri, and Maria Angela A. Meireles. 2019. "Comparative Analysis of Key Technologies for Cellulosic Ethanol Production from Brazilian Sugarcane Bagasse at a Commercial Scale." *Biofuels, Bioproducts and Biorefining* 13 (4): 994–1014. https://doi.org/10.1002/BBB.1990.

Costa, M. A. L., C. S. Farinas, and E. A. Miranda. 2018. "Ethanol Precipitation as a Downstream Processing Step for Concentration of Xylanases Produced by Submerged and Solid-State Fermentation." *Brazilian Journal of Chemical Engineering* 35 (2): 477–88. https://doi.org/10.1590/0104-6632.20180352s2016 0502.

Cutz, L., P. Haro, D. Santana, and F. Johnsson. 2016. "Assessment of Biomass Energy Sources and Technologies: The Case of Central America." *Renewable and Sustainable Energy Reviews* 58 (May): 1411–31. https://doi.org/10.1016/J.RSER. 2015.12.322.

DeAngelo, Julianne, Inês Azevedo, John Bistline, Leon Clarke, Gunnar Luderer, Edward Byers, and Steven J. Davis. 2021. "Energy Systems in Scenarios at Net-Zero CO2 Emissions." *Nature Communications 2021 12:1* 12 (1): 1–10. https://doi.org/10.1038/S41467-021-26356-Y.

Deuber, R. D. S., D. S. Fernandes, J. M. Bressanin, J. Watson, M. F. Chagas, A. Bonomi, L. V. Fregolente, and M. D. B. Watanabe. 2021. "Techno-Economic Assessment of HTL Integration to the Brazilian Sugarcane Industry: An Evaluation of Different Scenarios." *Industrial Crops and Products* 173. https://doi.org/10.1016/j.indcrop.2021.114139.

Díaz-Vázquez, D., D. A. Orozco-Nunnelly, C. Yebra-Montes, C. Senés-Guerrero, and M. S. Gradilla-Hernández. 2022. "Using Yeast Cultures to Valorize Tequila Vinasse Waste: An Example of a Circular Bioeconomy Approach in the Agro-Industrial Sector." *Biomass and Bioenergy* 161. https://doi.org/10.1016/j.biombioe.2022.106471.

Ellilä, Simo, Lucas Fonseca, Cristiane Uchima, Junio Cota, Gustavo Henrique Goldman, Markku Saloheimo, Vera Sacon, and Matti Siika-Aho. 2017. "Development of a Low-Cost Cellulase Production Process Using Trichoderma Reesei for Brazilian Biorefineries." *Biotechnology for Biofuels* 10: 30. https://doi.org/10.1186/s13068-017-0717-0.

Ferreira, G. F., L. F. Ríos Pinto, R. Maciel Filho, and L. V. Fregolente. 2021. "Effects of Cultivation Conditions on Chlorella Vulgaris and Desmodesmus Sp. Grown in Sugarcane Agro-Industry Residues." *Bioresource Technology* 342. https://doi.org/10.1016/j.biortech.2021.125949.

Finkler, Anelize Terezinha Jung, Alessandra Biz, Luana Oliveira Pitol, Bruna Schweitzer Medina, Henrique Luithardt, Luiz Fernando de Lima Luz, Nadia Krieger, and David Alexander Mitchell. 2017. "Intermittent Agitation Contributes to Uniformity across the Bed during Pectinase Production by Aspergillus Niger Grown in Solid-State Fermentation in a Pilot-Scale Packed-Bed Bioreactor." *Biochemical Engineering Journal* 121 (May): 1–12. https://doi.org/10.1016/J.BEJ.2017.01.011.

Fuess, Lucas Tadeu, Moacir Messias de Araújo Júnior, Marcelo Loureiro Garcia, and Marcelo Zaiat. 2017. "Designing Full-Scale Biodigestion Plants for the Treatment of Vinasse in Sugarcane Biorefineries: How Phase Separation and Alkalinization Impact Biogas and Electricity Production Costs?" *Chemical Engineering Research and Design* 119 (March): 209–20. https://doi.org/10.1016/J.CHERD.2017.01.023.

Fuess, Lucas Tadeu, Marcelo Loureiro Garcia, and Marcelo Zaiat. 2018. "Seasonal Characterization of Sugarcane Vinasse: Assessing Environmental Impacts from Fertirrigation and the Bioenergy Recovery Potential through Biodigestion." *Science of The Total Environment* 634 (September): 29–40. https://doi.org/10.1016/J.SCITOTENV.2018.03.326.

Fuess, Lucas Tadeu, Bruno Colling Klein, Mateus Ferreira Chagas, Mylene Cristina Alves Ferreira Rezende, Marcelo Loureiro Garcia, Antonio Bonomi, and Marcelo Zaiat. 2018. "Diversifying the Technological Strategies for Recovering Bioenergy from the Two-Phase Anaerobic Digestion of Sugarcane Vinasse: An Integrated Techno-Economic and Environmental Approach." *Renewable Energy* 122 (July): 674–87. https://doi.org/10.1016/J.RENENE.2018.02.003.

Fuess, Lucas Tadeu, and Marcelo Zaiat. 2018. "Economics of Anaerobic Digestion for Processing Sugarcane Vinasse: Applying Sensitivity Analysis to Increase Process Profitability in Diversified Biogas Applications." *Process Safety and Environmental Protection* 115 (April): 27–37. https://doi.org/10.1016/J.PSEP.2017.08.007.

García, Carlos A., Ramiro Betancourt, and Carlos A. Cardona. 2017. "Stand-Alone and Biorefinery Pathways to Produce Hydrogen through Gasification and Dark Fermentation Using Pinus Patula." *Journal of Environmental Management* 203 (December): 695–703. https://doi.org/10.1016/J.JENVMAN.2016.04.001.

Gómez, J. A., P. Berni, L. G. Matallana, Ó. J. Sánchez, J. A. Teixeira, and C. Nobre. 2022. "Towards a Biorefinery Processing Waste from Plantain Agro–Industry: Process Development for the Production of an Isomalto–Oligosaccharide Syrup from Rejected Unripe Plantain Fruits." *Food and Bioproducts Processing* 133: 100–118. https://doi.org/10.1016/j.fbp.2022.03.005.

Gómez, J. A., C. Nobre, J. A. Teixeira, and Ó. J. Sánchez. 2022. "Towards a Biorefinery Processing Waste from Plantain Agro-Industry: Assessment of the Production of Dairy Cattle Feed through Process Simulation." *Biosystems Engineering* 217: 131–49. https://doi.org/10.1016/j.biosystemseng.2022.03.008.

González, G., N. V. Ehman, Y. S. Aguerre, S. Henríquez-Gallegos, A. P. N. da Fonte, G. I. B. Muniz, M. Pereira, et al. 2022. "Quality of Microfibrillated Cellulose Produced from Unbleached Pine Sawdust Pulp as an Environmentally Friendly Source." *Waste and Biomass Valorization* 13 (3): 1609–26. https://doi.org/10.1007/s12649-021-01615-7.

Grangeia, Carolina, Luan Santos, and Lira Luz Benites Lazaro. 2022. "The Brazilian Biofuel Policy (RenovaBio) and Its Uncertainties: An Assessment of Technical, Socioeconomic and Institutional Aspects." *Energy Conversion and Management: X* 13 (January): 100156. https://doi.org/10.1016/J.ECMX.2021.100156.

Guimarães, H. R., J. M. Bressanin, I. L. Motta, M. F. Chagas, A. Bonomi, R. M. Filho, and M. D. B. Watanabe. 2022. "Techno-Economic and Environmental Assessments of Thermochemical Routes Integrated into the Brazilian Sugarcane Industry for the Production of Renewable Jet Fuel." *Chemical Engineering Transactions* 92: 115–20. https://doi.org/10.3303/CET2292020.

Imlauer Vedoya, C. M., M. C. Area, N. Raffaeli, and F. E. Felissia. 2022. "Study on Soda–Ethanol Delignification of Pine Sawdust for a Biorefinery." *Sustainability (Switzerland)* 14 (11). https://doi.org/10.3390/su14116660.

Katayama, E., N. A. Rodrigues, S. Bilatto, F. P. Casciatori, and C. S. Farinas. 2022. "Nanocellulose Isolation Using a Thermostable Endoglucanase-Rich Cocktail from Myceliophthora Thermophila Cultivated in a Multilayer Packed-Bed Bioreactor." *Biomass Conversion and Biorefinery*. https://doi.org/10.1007/s13399-022-02977-1.

Klein, Bruno Colling, Mateus Ferreira Chagas, Tassia Lopes Junqueira, Mylene Cristina Alves Ferreira Rezende, Terezinha de Fátima Cardoso, Otavio Cavalett, and Antonio Bonomi. 2018. "Techno-Economic and Environmental Assessment of Renewable Jet Fuel Production in Integrated Brazilian Sugarcane Biorefineries." *Applied Energy* 209 (January): 290–305. https://doi.org/10.1016/J.APENERGY.2017.10.079.

Klein, Bruno Colling, Mateus Ferreira Chagas, Marcos Djun Barbosa Watanabe, Antonio Bonomi, and Rubens Maciel Filho. 2019. "Low Carbon Biofuels and the New Brazilian National Biofuel Policy (RenovaBio): A Case Study for Sugarcane Mills and Integrated Sugarcane-Microalgae Biorefineries." *Renewable and Sustainable Energy Reviews* 115 (November): 109365. https://doi.org/10.1016/J.RSER.2019. 109365.

López-Díaz, Dulce Celeste, Luis Fernando Lira-Barragán, Eusiel Rubio-Castro, Medardo Serna-González, Mahmoud M. El-Halwagi, and José María Ponce-Ortega. 2018. "Optimization of Biofuels Production via a Water–Energy–Food Nexus Framework." *Clean Technologies and Environmental Policy* 20 (7): 1443–66. https://doi.org/10.1007/S10098-017-1395-0/FIGURES/15.

Machin Ferrero, L. M., and F. D. Mele. 2022. "Optimal Design of Sugar-Cane-Based Biorefinery Networks in Argentina." *ACS Sustainable Chemistry and Engineering*. https://doi.org/10.1021/acssuschemeng.2c00921.

Mariana, Ortiz Sanchez, Cardona Alzate, and Carlos Ariel. 2021. "Comparative Environmental Life Cycle Assessment of Orange Peel Waste in Present Productive Chains." *Journal of Cleaner Production* 322 (November): 128814. https://doi.org/10.1016/J.JCLEPRO.2021.128814.

Martins, L. B., J. Soares, W. B. da Silveira, R. C. S. Sousa, and M. A. Martins. 2021. "Dilute Sulfuric Acid Hydrolysis of Chlorella Vulgaris Biomass Improves the Multistage Liquid-Liquid Extraction of Lipids." *Biomass Conversion and Biorefinery* 11 (6): 2485–97. https://doi.org/10.1007/s13399-020-00661-w.

Mello, Matheus Lima de, Marco Aurelio Sperança, and Fabíola Manhas Verbi Pereira. 2022. "Mineral Scale Formed During the Evaporation in Biorefineries — Characterization Using X-Ray Fluorescence and Exploratory Analysis." *Food Analytical Methods* 1 (April): 1–4. https://doi.org/10.1007/S12161-022-02308-0/FIGURES/3.

Mendieta, C. M., J. Kruyeniski, F. E. Felissia, and M. C. Area. 2022. "Modelling of the Simultaneous Saccharification and Fermentation for a Pine Sawdust Biorefinery." *Fermentation* 8 (3). https://doi.org/10.3390/fermentation8030130.

Menon, Vishnu, and Mala Rao. 2012. "Trends in Bioconversion of Lignocellulose: Biofuels, Platform Chemicals & Biorefinery Concept." *Progress in Energy and Combustion Science* 38 (4): 522–50. https://doi.org/10.1016/J.PECS.2012.02.002.

Michelon, W., M. L. B. da Silva, A. Matthiensen, C. J. de Andrade, L. M. de Andrade, and H. M. Soares. 2022. "Amino Acids, Fatty Acids, and Peptides in Microalgae Biomass Harvested from Phycoremediation of Swine Wastewaters." *Biomass Conversion and Biorefinery* 12 (3): 869–80. https://doi.org/10.1007/s13399-020-01263-2.

Montoya, G., M. I. Gutierrez, J. D. Giraldo, L. D. Jaramillo, J. Ruiz-Sandoval, S. Orozco, F. Orozco, J. Ward, G. Rojas, and M. F. Villegas-Torres. 2021. "Sustainable Sugarcane Vinasse Biorefinement for Trans-Aconitic Acid-Based Biopolymer Synthesis and Bioenergy Generation." *Bioresource Technology Reports* 15. https://doi.org/10.1016/j.biteb.2021.100786.

Morales Chavez, M. M., Y. Costa, and W. Sarache. 2021. "A Three-Objective Stochastic Location-Inventory-Routing Model for Agricultural Waste-Based Biofuel Supply Chain." *Computers and Industrial Engineering* 162. https://doi.org/10.1016/j.cie.2021.107759.

Morales-Rodriguez, R., J. D. Ponce-Rocha, F. I. Gómez-Castro, E. Sánchez-Ramírez, J. G. Segovia-Hernández, A. Sánchez, and D. Rodriguez-Gomez. 2021. "Acetone, Butanol, Ethanol and, Xylitol Production Through a Biorefinery Platform: An Experimental & Simulation Approach." *Waste and Biomass Valorization* 12 (9): 4915–30. https://doi.org/10.1007/s12649-020-01327-4.

Morreeuw, Z. P., L. J. Ríos-González, C. Salinas-Salazar, E. M. Melchor-Martínez, J. A. Ascacio-Valdés, R. Parra-Saldívar, H. M. N. Iqbal, and A. G. Reyes. 2021. "Early Optimization Stages of Agave Lechuguilla Bagasse Processing toward Biorefinement: Drying Procedure and Enzymatic Hydrolysis for Flavonoid Extraction." *Molecules* 26 (23). https://doi.org/10.3390/molecules26237292.

Ortiz-Sanchez, M., J. C. Solarte-Toro, C. E. Orrego-Alzate, C. D. Acosta-Medina, and C. A. Cardona-Alzate. 2021. "Integral Use of Orange Peel Waste through the Biorefinery Concept: An Experimental, Technical, Energy, and Economic Assessment." *Biomass Conversion and Biorefinery* 11 (2): 645–59. https://doi.org/ 10.1007/S13399-020-00627-Y/FIGURES/3.

Pedrozo, H. A., A. I. Casoni, F. D. Ramos, V. Estrada, and M. S. Diaz. 2022. "Simultaneous Design of Macroalgae-Based Integrated Biorefineries and Their Heat Exchanger Network." *Computers and Chemical Engineering* 164. https://doi.org/10.1016/j.compchemeng.2022.107885.

Peña-Jurado, E., Samuel Pérez-Vega, Francisco Javier Zavala-Díaz de la Serna, Ildebrando Pérez-Reyes, Nestor Gutiérrez-Méndez, José Vazquez-Castillo, and Ivan Salmerón.

2019. "Production of Poly (3-Hydroxybutyrate) from a Dairy Industry Wastewater Using Bacillus Subtilis EPAH18: Bioprocess Development and Simulation." *Biochemical Engineering Journal* 151 (November): 107324. https://doi.org/10.1016/J.BEJ.2019.107324.

Pereira, B., W.F. Marcondes, W. Carvalho, and V. Arantes. 2021. "High Yield Biorefinery Products from Sugarcane Bagasse: Prebiotic Xylooligosaccharides, Cellulosic Ethanol, Cellulose Nanofibrils and Lignin Nanoparticles." *Bioresource Technology* 342. https://doi.org/10.1016/j.biortech.2021.125970.

Perez, C. L., L. P. R. D. C. Pereira, T. S. Milessi, J. P. Sandri, M. Demeke, M. R. Foulquié-Moreno, J. M. Thevelein, and T. C. Zangirolami. 2022. "Towards a Practical Industrial 2G Ethanol Production Process Based on Immobilized Recombinant S. Cerevisiae: Medium and Strain Selection for Robust Integrated Fixed-Bed Reactor Operation." *Renewable Energy* 185: 363–75. https://doi.org/ 10.1016/j.renene.2021.12.044.

Pinto, A. S. S., A. M. Elias, F. F. Furlan, M. P. A. Ribeiro, R. C. Giordano, and C. S. Farinas. 2021. "Techno-Economic Feasibility of Biomass Washing in 1G2G Sugarcane Biorefineries." *Bioenergy Research* 14 (4): 1253–64. https://doi.org/10.1007/s12155-020-10230-8.

Rachid-Casnati, C., F. Resquin, and L. Carrasco-Letelier. 2021. "Availability and Environmental Performance of Wood for a Second-Generation Biorefinery." *Forests* 12 (11). https://doi.org/10.3390/F12111609.

Ramos, L. R., G. Lovato, J. A. D. Rodrigues, and E. L. Silva. 2022. "Scale-up and Energy Estimations of Single- and Two-Stage Vinasse Anaerobic Digestion Systems for Hydrogen and Methane Production." *Journal of Cleaner Production* 349. https://doi.org/10.1016/j.jclepro.2022.131459.

Rangel, C., J. Sastoque, J. Calderón, J. Gracia, I. Cabeza, S. Villamizar, and P. Acevedo. 2022. "Pilot-Scale Assessment of Biohydrogen and Volatile Fatty Acids Production via Dark Fermentation of Residual Biomass." *Chemical Engineering Transactions* 92: 61–66. https://doi.org/10.3303/CET2292011.

Restrepo-Serna, D. L., J. C. Solarte-Toro, and C. A. Cardona-Alzate. 2022. "A Biorefinery Approach for an Integral Valorisation of Avocado Peel and Seeds through Supercritical Fluids." *Waste and Biomass Valorization*. https://doi.org/10.1007/s12649-022-01829-3.

Rodríguez, F., E. Aguilar-Garnica, A. Santiago-Toribio, and A. Sánchez. 2022. "Polysaccharides Release in a Laboratory-Scale Batch Hydrothermal Pretreatment of Wheat Straw under Rigorous Isothermal Operation." *Molecules* 27 (1). https://doi.org/10.3390/molecules27010026.

Rosales-Calderon, Oscar, and Valdeir Arantes. 2019. "A Review on Commercial-Scale High-Value Products That Can Be Produced alongside Cellulosic Ethanol." *Biotechnology for Biofuels* 12 (1): 1–58. https://doi.org/10.1186/S13068-019-1529-1/FIGURES/6.

Sánchez Muñoz, Salvador, Thércia Rocha Balbino, Edith Mier Alba, Fernanda Gonçalves Barbosa, Fernando Tonet de Pier, Alexandra Lazuroz Moura de Almeida, Ana Helena Balan Zilla, et al. 2022. "Surfactants in Biorefineries: Role, Challenges & Perspectives." *Bioresource Technology* 345 (February): 126477. https://doi.org/10.1016/J.BIORTECH.2021.126477.

Santana, Hugo, Carolina R. Cereijo, Valérya C. Teles, Rodrigo C. Nascimento, Maiara S. Fernandes, Patrícia Brunale, Raquel C. Campanha, et al. 2017. "Microalgae Cultivation in Sugarcane Vinasse: Selection, Growth and Biochemical Characterization." *Bioresource Technology* 228 (March): 133–40. https://doi.org/10.1016/J.BIORTECH.2016.12.075.

Schneider, Willian Daniel Hahn, Roselei Claudete Fontana, Simone Mendonça, Félix Gonçalves de Siqueira, Aldo José Pinheiro Dillon, and Marli Camassola. 2018. "High Level Production of Laccases and Peroxidases from the Newly Isolated White-Rot Basidiomycete Marasmiellus Palmivorus VE111 in a Stirred-Tank Bioreactor in Response to Different Carbon and Nitrogen Sources." *Process Biochemistry* 69 (June): 1–11. https://doi.org/10.1016/J.PROCBIO.2018.03.005.

Serna-Loaiza, Sebastián, Estefanny Carmona-Garcia, and Carlos A. Cardona. 2018. "Potential Raw Materials for Biorefineries to Ensure Food Security: The Cocoyam Case." *Industrial Crops and Products* 126 (December): 92–102. https://doi.org/10.1016/J.INDCROP.2018.10.005.

Sganzerla, W. G., D. Lachos-Perez, L. S. Buller, G. L. Zabot, and T. Forster-Carneiro. 2022. "Cost Analysis of Subcritical Water Pretreatment of Sugarcane Straw and Bagasse for Second-Generation Bioethanol Production: A Case Study in a Sugarcane Mill." *Biofuels, Bioproducts and Biorefining* 16 (2): 435–50. https://doi.org/10.1002/bbb.2332.

Silva-Martínez, Rodolfo Daniel, Alessandro Sanches-Pereira, Willington Ortiz, Maria Fernanda Gómez Galindo, and Suani Teixeira Coelho. 2020. "The State-of-the-Art of Organic Waste to Energy in Latin America and the Caribbean: Challenges and Opportunities." *Renewable Energy* 156 (August): 509–25. https://doi.org/10.1016/J.RENENE.2020.04.056.

Solarte-Toro, J. C., M. Ortiz-Sanchez, and C. A. Cardona Alzate. 2022. "Environmental Life Cycle Assessment (E-LCA) and Social Impact Assessment (SIA) of Small-Scale Biorefineries Implemented in Rural Zones: The Avocado (Persea Americana Var. Americana) Case in Colombia." *Environmental Science and Pollution Research*. https://doi.org/10.1007/s11356-022-20857-z.

Solarte-Toro, J. C., M. Ortiz-Sanchez, D. L. Restrepo-Serna, P. Peroza Piñeres, A. Pérez Cordero, and C. A. Cardona Alzate. 2021. "Influence of Products Portfolio and Process Contextualization on the Economic Performance of Small- and Large-Scale Avocado Biorefineries." *Bioresource Technology* 342. https://doi.org/10.1016/j.biortech.2021.126060.

Solarte-Toro, Juan Camilo, Mariana Ortiz-Sanchez, and Carlos Ariel Cardona Alzate. 2022. "Environmental Life Cycle Assessment (E-LCA) and Social Impact Assessment (SIA) of Small-Scale Biorefineries Implemented in Rural Zones: The Avocado (Persea Americana Var. Americana) Case in Colombia." *Environmental Science and Pollution Research*, June, 1–19. https://doi.org/10.1007/S11356-022-20857-Z/TABLES/11.

Sosa-Martínez, J., N. Balagurusamy, J. R. Benavente-Valdés, J. Montañez, and L. Morales-Oyervides. 2021. "Process Performance Improvement for the Simultaneous Production of Ligninolytic Enzymes in Solid Culture Using Agricultural Wastes through the Taguchi Method." *Journal of Environmental Management* 293. https://doi.org/10.1016/j.jenvman.2021.112966.

Sousa, M. H. de, A. S. F. da Silva, R. C. Correia, N. P. Leite, C. E. G., Bueno, R. L. dos Santos Pinheiro, J. S. de Santana, et al. 2022. "Valorizing Municipal Organic Waste to Produce Biodiesel, Biogas, Organic Fertilizer, and Value-Added Chemicals: An Integrated Biorefinery Approach." *Biomass Conversion and Biorefinery* 12 (3): 827–41. https://doi.org/10.1007/s13399-020-01252-5.

Tessari, L. F. A., C. R. Soccol, C. Rodrigues, E. G., González, V. O. A. Tanobe, P. C. S. Kirnev, and J. C. de Carvalho. 2022. "Development of a Culture Medium for Microalgae Production Based on Minimal Processing of Oil Palm Biomass Ash." *Fermentation* 8 (2). https://doi.org/10.3390/fermentation8020055.

Uchôa, P. Z., R. C. T. Porto, R. Battisti, C. Marangoni, N. Sellin, and O. Souza. 2021. "Ethanol from Residual Biomass of Banana Harvest and Commercialization: A Three-Waste Simultaneous Fermentation Approach and a Logistic-Economic Assessment of the Process Scaling-up towards a Sustainable Biorefinery in Brazil." *Industrial Crops and Products* 174. https://doi.org/10.1016/j.indcrop.2021.114170.

Valentim Bastos, J. B., J. G. Soares Santos Maia, S. Borschiver, A. Szklo, and A. R. Secchi. 2022. "Addressing Scale and Seasonality in the Design of Sugarcane to Ethylene Glycol Biorefineries." *Journal of Cleaner Production* 337. https://doi.org/10.1016/j.jclepro.2022.130585.

Volpi, M. P. C., L. B. Brenelli, G. Mockaitis, S. C. Rabelo, T. T. Franco, and B. S. Moraes. 2022. "Use of Lignocellulosic Residue from Second-Generation Ethanol Production to Enhance Methane Production Through Co-Digestion." *Bioenergy Research* 15 (1): 602–16. https://doi.org/10.1007/s12155-021-10293-1.

Volpi, M. P. C., A. D. N. F. Junior, T. T. Franco, and B. S. Moraes. 2021. "Operational and Biochemical Aspects of Co-Digestion (Co-AD) from Sugarcane Vinasse, Filter Cake, and Deacetylation Liquor." *Applied Microbiology and Biotechnology* 105 (23): 8969–87. https://doi.org/10.1007/s00253-021-11635-x.

Watanabe, Marcos D. B., Lucas G. Pereira, Mateus F. Chagas, Marcelo Pereira da Cunha, Charles D. F. Jesus, Alexandre Souza, Elmer C. Rivera, R. Maciel Filho, Otávio Cavalett, and Antonio Bonomi. 2016. "*Sustainability Assessment Methodologies*," 155–88. https://doi.org/10.1007/978-3-319-26045-7_6.

Zuorro, A., K. A. Moreno-Sader, and Á. D. González-Delgado. 2021. "Evaluating the Feasibility of a Pilot-Scale Shrimp Biorefinery via Techno-Economic Analysis." *Journal of Cleaner Production* 320. https://doi.org/10.1016/j.jclepro.2021.128740.

Chapter 4

A Spatially Explicit Assessment of a Forestry Biorefinery for the Recovery of Bio-Based Materials

W. G. Sganzerla[1], C. W. S. Romero[2], L. S. Buller[1], M. D. Berni[3], R. A. Lamparelli[3], T. Forster-Carneiro[1] and T. T. Franco[4]

[1]School of Food Engineering (FEA),
University of Campinas (UNICAMP), Campinas, SP, Brazil
[2]School of Agricultural Engineering (FEAGRI),
University of Campinas (UNICAMP), Campinas, SP, Brazil
[3]Center for Energy Planning,
University of Campinas (UNICAMP), Campinas, SP, Brazil
[4]School of Chemical Engineering (FEQ),
University of Campinas (UNICAMP), Campinas, SP, Brazil

Abstract

The biorefinery concept emerged as an alternative to decrease the greenhouse gases emissions by the industrial section while producing bioenergy and value-added products. This chapter evaluated a critical analysis of the residues derived from eucalyptus harvest to support decision-making for a biorefinery implementation in the Administrative Region of Campinas (ARC), São Paulo State, Brazil. Brazilian official data for forestry extraction were collected for the eucalyptus production areas of the municipalities of ARC. RapidEye images were used to process, identify, and classify the areas occupied by eucalyptus and the chemical industries, aiming to compose a georeferenced database. Although most of the developments have been focused on processing technologies, it is critical to define the specific biobased products that

In: The Future of Biorefineries
Editor: Waldemar Nyström
ISBN: 979-8-88697-524-6

should be produced considering market drivers such as demand, production flexibility, logistics, and supply chain configuration. The conversion of lignocellulosic materials present in forestry residues could produce several bio-based products after a pretreatment step. These products could replace fossil-based raw materials for the chemical industry, such as polyurethane, lignin, tannins, furfural, xylose, xylitol, polyphenols, and lubricants. It was possible to observe the existence of promising potential for the use of eucalyptus residues to produce a wide variety of chemicals. In the case of biofuels, the use of residual biomass is a primary factor for the transition to a viable and more renewable energy chain since eucalyptus residues are still little or not explored.

Keywords: biorefinery, eucalyptus residues, bioenergy, biofuels, bio-based products

1. Introduction

Lignocellulosic materials, including forestry residues, are a significant part of natural carbohydrates worldwide (Wang et al., 2020). The conversion of these materials in biorefineries comes up with an attractive and promising alternative to producing bio-products (Ajao et al., 2021). In addition to reducing greenhouse gas emissions, it does not compete for food production systems while it does not lead to increasing cropland areas (Popp et al., 2014). Worldwide, more than 90% of petroleum products are expected to be produced from renewable raw materials, and half of the pharmaceutical market could be bio-based by the year 2030 (Clauser et al., 2021). Moreover, the forest plantation includes several hardwoods (eucalyptus, birch, others) and softwoods (pines, spruces, larch, others), which the global residues production can reach 224×10^6 t y^{-1} (Millati et al., 2019). Eucalyptus residues composition presents cellulose, hemicellulose, and lignin contents varying according to the species, growth conditions, harvesting age, and plant part (Parajuli et al., 2015). The lignin can be disjoined from the forestry fibers and the cellulosic fraction by thermochemical and/or biochemical pretreatment technologies (Stafford et al., 2020). The waste-derived from industrial eucalyptus processing for paper and cellulose production is mainly repurposed as biofuel (black liquor) for electric energy generation. For the residues from eucalyptus harvesting (bark and branches), it is possible to produce steam in boilers and bio-based products (Pinto et al., 2013).

Bio-based product's potential application could supply not only the own cellulosic industry but also other industrial segments such as wood for construction, automotive (filters), chemical, pharmaceutical, and food industries (Dornelles et al., 2021). The bio-products produced in forestry biorefineries could replace fossil-based raw materials, such as polyurethane lignin, furfural, xylose, xylitol, tannins, polyphenols, and lubricants (Orejuela-Escobar et al., 2021). Notwithstanding, bio-based products can play a significant role in the circular economy transition towards economic decarbonization and are indispensable components in the *portfolio* of low carbon technologies required to limit the current climate change like the global warming (Junginger et al., 2019).

Therefore, a critical analysis of the residues derived from eucalyptus harvesting was presented to support decision-making processes for a biorefinery implementation in the Administrative Region of Campinas (ARC), São Paulo State, Brazil.

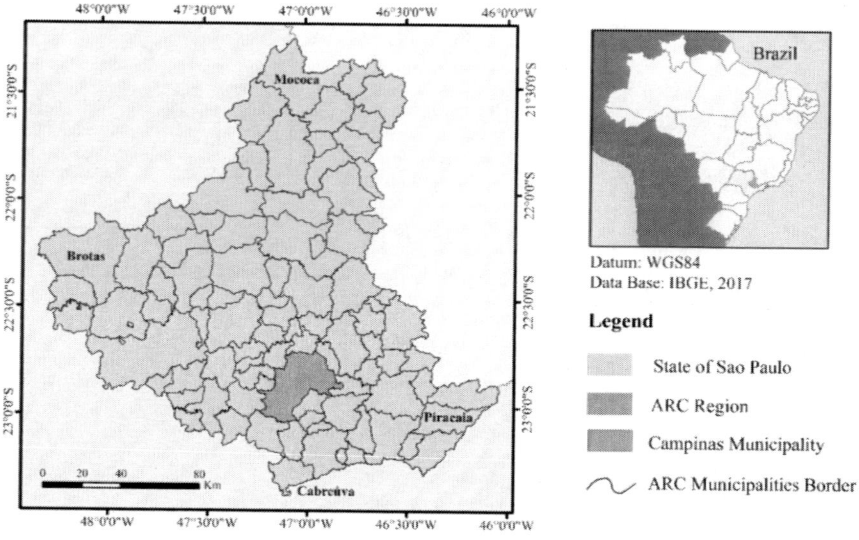

Figure 1. ARC geographical location.

2. Methodology

2.1. Geographic Scope of the Study

From the inventory of cultivated eucalyptus barks (the main residues generated) in ARC, ethanol, soluble extracts (i.e., tannins for wood adhesives), and polyurethane lignin industrial production were estimated. A data collection from the ARC chemical industries profile was performed to select the bio-products of interest, mainly based on the market demand. ARC accounts for approximately 11% of São Paulo State territory, occupying 27,093 km² divided into 92 municipalities, housing more than 6.9 million inhabitants (Figure 1). ARC Gross Domestic Product (GDP) is approximately USD 93.8 billion (18% of São Paulo State's total GDP). Beyond, the ARC's chemical industries accounted for 204 companies in 2015.

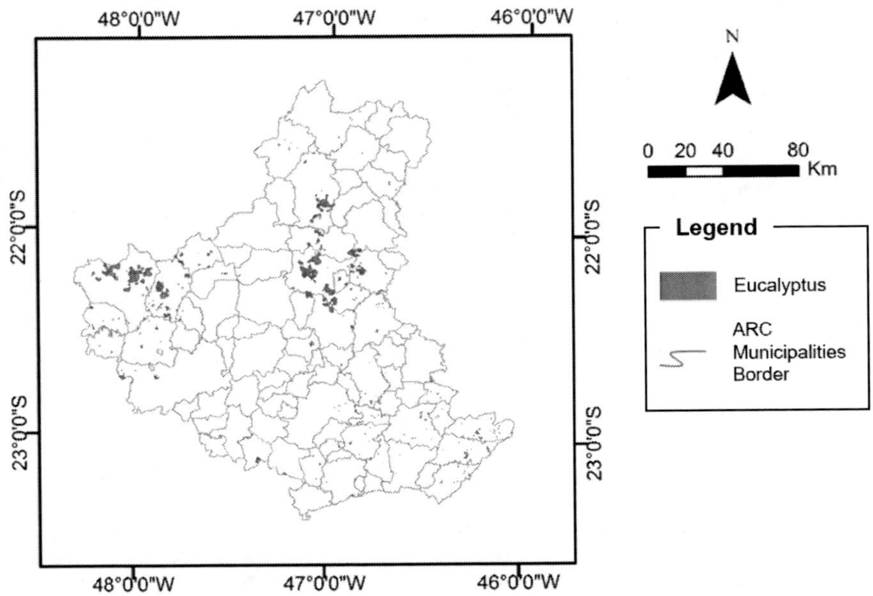

Figure 2. Eucalyptus production areas.

2.2. Database and Information Georeferencing

The database for the eucalyptus production areas was collected from the Le Maire et al., (2014) study (Figure 2). The authors mapped the eucalyptus areas in Brazil using a binary classification method Normalized Difference Vegetation Index (NDVI), based on MODIS sensor images (250 meters of spatial resolution). The update of these areas was identified by visual analysis of RapidEye images (cloud-free). The referred images are characterized as follows: spatial resolution of 5 meters; radiometric of 12 bits; spectral of 5 bands (blue, green, red, red-edge, and near-infrared); and temporal of 5.5 days at Nadir acquiring. The colored composition NIR-R, RED-G, BLUE-B was used for the manual area vectoring. Because the eucalyptus harvesting cycle is based on 5.5 to 7 years, temporal analysis is mandatory to detect land cover changes during seasonal periods and avoid misconstruction of land use/image classification. All chemical industries were spatialized from their geographical coordinates established on the municipal level (Figure 3).

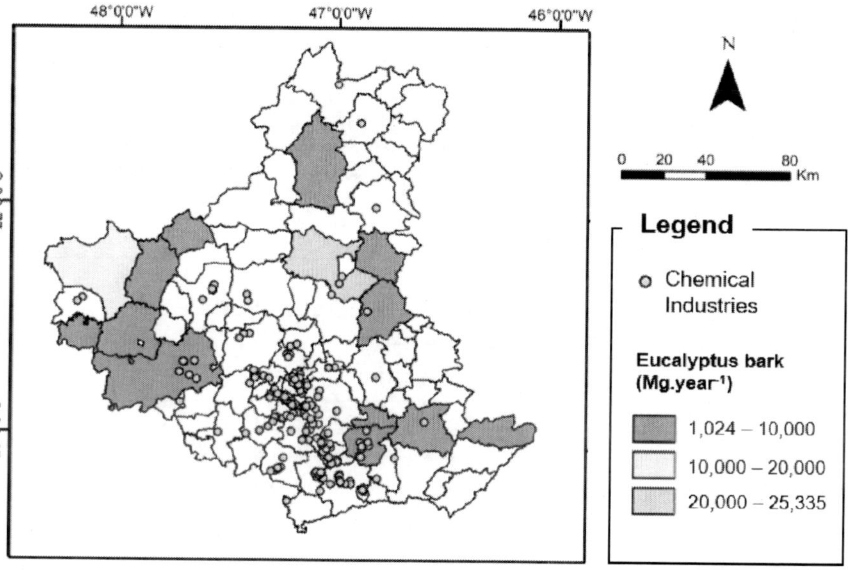

Figure 3. Chemical industries and eucalyptus bark availability in ARC.

2.3. Eucalyptus Barks Availability

The estimation for eucalyptus barks availability was obtained from the production of the log, according to Eq. (1) (Foelkel, 2016; Pincelli et al., 2017).

$$BA = LP * 0.14 * 0.325 * 0.9 \quad (1)$$

where: BA, barks availability ($t\ y^{-1}$); LP, logs production ($m^3\ y^{-1}$); 0.14 is the bark's average content for eucalyptus; 0.325 is the bark's average specific weight in t m^{-3}; and 0.9 corresponds to the dry mass for 10% humidity in eucalyptus bark.

For the simulation, the fifteen (15) major municipalities producing eucalyptus logs were considered (Table 1). Mogi Guaçu and Brotas reach 41% of the total harvesting in the study area from those municipalities.

Table 1. Availability of eucalyptus barks (base year, 2018)

Municipalities	Logs production (m^3 y^{-1})	Bark (dry basis) (t y^{-1})
Analândia	96,500	3,952
Bragança Paulista	135,000	5,528
Brotas	299,551	12,267
Casa Branca	215,000	8,804
Espírito Santo do Pinhal	47,000	1,925
Itapira	46,588	1,908
Itatiba	26,500	1,085
Itirapina	242,880	9,946
Joanópolis	182,455	7,472
Mogi Guaçu	618,684	25,335
Morungaba	67,000	2,744
Piracaia	25,000	1,024
Piracicaba	124,500	5,098
Santa Maria da Serra	54,350	2,226
São Pedro	79,820	3,269
Total	2,260,828	92,581

2.4. Production of Bio-Based Products

From the estimates of eucalyptus barks availability, the potential production of first (E1G) and second-generation ethanol (E2G) can be calculated (Eq. 2), according to the following yields (Romero et al., 2019): 106 L E1G t^{-1} eucalyptus bark; and 94 L E2G t^{-1} residual solid biomass.

$$Ethanol = BA * 106 + DCE * 0.75 * 94 \quad (2)$$

where: BA, barks availability (t y^{-1}); 106 is the yield of E1G; 0.75 refers that approximately 75% of the original mass of bark used for E1G; and 94 is the yield of E2G.

The extraction of tannins from eucalyptus barks can be done with hot water and the addition of 4.5% sodium sulphite, for three hours, at temperatures of 70 – 100 °C. In the production of wood adhesive, tannins react with acetic acid and sodium sulphite to reduce viscosity. For every 30 kg of eucalyptus barks, 3.2 kg of tannins can be obtained (Dababi et al., 2020). Beyond, 17% of eucalyptus barks could be extracted in the form of tannins, and every 50 g of tannin could produce 100 mL of wood adhesive, reaching the ratio 214 L of wood adhesive per ton of bark (Vital et al., 2004). From these previous data, the potential to generate wood adhesive in the ARC was estimated according to Eq. (3).

$$Wood\ adhesive = BA * 0.17 * 214 \quad (3)$$

where: BA, barks availability (t^{-1}); 0.17 is the proportion of eucalyptus bark that can be extracted in the form of tannins; and 214 is the ratio of wood adhesive per ton of bark ($L\ t^{-1}$).

The potential generation of polyurethane lignin was estimated according to Eq. (4). The lignin content in eucalyptus barks can vary from 12 to 20% (Andrade et al., 2021). For the estimates, 16% of lignin extraction from eucalyptus barks was adopted, and the assumption that 10% goes to the production of polyurethane lignin was done.

$$Polyurethane\ lignin = BA * 0.16 * 0.10 \quad (4)$$

where: BA, barks availability (t^{-1}); 0.16 is the average lignin content; and 0.10 in the amount of eucalyptus bark that goes into the production of polyurethane lignin.

3. Result and Discussion

The fractionation of lignocellulosic waste is a strategy to facilitate the subsequent processes of converting cellulose into fermentable sugars which can be destined for ethanol production or other products of commercial interest (Ajao et al., 2021). The biorefinery technological arrangement employs the biomass components that can be separated into different fractions, such as sugars (from polysaccharides) and phenolic compounds (from lignin), obtaining value-added bio-products (Clauser et al., 2021). In this study, the proposed biorefinery could produce ethanol, soluble extracts, wood adhesive, and polyurethane lignin (Figure 4).

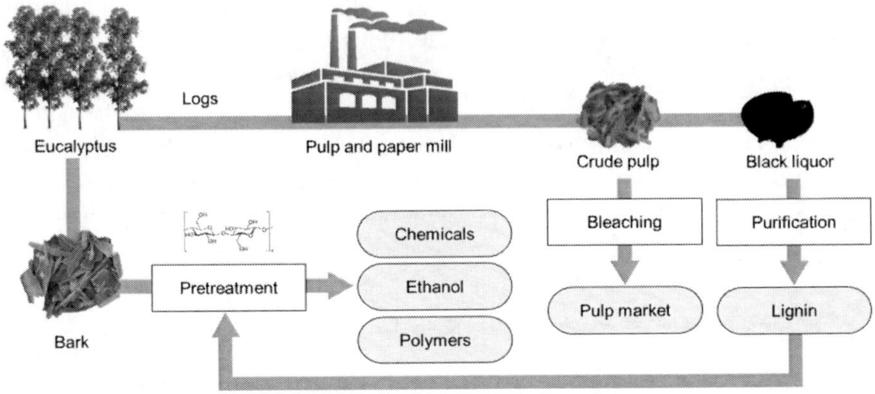

Figure 4. Scheme of the proposed biorefinery to the valorization of eucalyptus barks.

From the proposed biorefinery, the production of E1G and E2G could be estimated according to the ethanol yield from eucalyptus barks, where approximately 75% of the original bark mass is used up in the E1G production. The amount of hydrated ethanol produced using 92,581 t of ARC's eucalyptus barks accounts for 9,814 m^3 and 6,527 m^3, respectively, of E1G and E2G. Considering São Paulo State's consumption of hydrated ethanol in 2018 of 9,956,761 m^3 (UNICA, 2020), the potential production of solely 15 municipalities would correspond to 0.17% of the current consumption. Specifically, concerning the consumption of the 15 municipalities of the ARC, the production of ethanol would supply 5.3% of the region. On the other hand, the potential production of hydrated ethanol (16,341 m^3) would correspond to 0.23% of the hydrated ethanol produced in the 2017/2018 sugarcane harvest in the State of São Paulo (7,260,000 m^3) (SIMA, 2020). Other technological routes available for biorefinery implementation are the production of tannins and soluble extracts. Tannins are basic raw materials for wood adhesive production. After the extraction of the water-soluble compounds, lignin is extracted, and, lastly, the lignin can be destined for the production of polyurethane lignin. In the proposed biorefinery for the ARC, the theoretical production of wood adhesive and polyurethane lignin can be estimated at approximately 787 and 1,481 t y^{-1}, respectively. The optimal location for a biorefinery could be determined from the potential bio-products estimates. The study of optimal location is a vital decision-making tool related to upstream and downstream supply-chain structure. Such decisions involve determining the number, location, production capacity, and occupation of the facilities, including nodal points in the distribution network to the final consumer market (Luhas et al., 2021). Distribution costs are typically the most relevant, and the trade-off between them and the distance to the potential consumer market could be established from the spatial analysis of the consumer market. From the premise that a single biorefinery could promote economies of scope and scale to meet the demand, as well as the optimization of installation and inventory costs, an analysis of logistics costs, especially transportation, could support the decision-making toward the transition to a bio-based economy (AliAkbari et al., 2021). The presented approach could be considered a first step to achieving a sustainable transition.

Conclusion

In Brazil, a continental country area, the biomass economic potential is a topic of great interest due to the large generation of agro-industrial waste. However, few studies focus exclusively on agro-industrial waste, except for projects related to second-generation ethanol production and industrial effluents recovery for energy purposes, such as sugarcane vinasse. In the case of biofuels, the use of residual biomass is a primary factor for the transition to a viable and more renewable energy chain. Barks derived from eucalyptus crops are still little or not explored. In this initial study, it was possible to observe the existence of promising potential for the use of eucalyptus residues to produce a wide variety of chemicals. The low density of bark biomass in the agricultural eucalyptus phase generates undesirable logistical costs. However, studies like this can support the most appropriate location for a possible biorefinery.

Acknowledgments

The authors acknowledge the financial support from FAEPEX/UNICAMP/EDUCORP, São Paulo Research Foundation (FAPESP) (grants 2015/50612-8, 2018/14938-4, 2019/26925-7, and 2020/10323-5), Coordination for the Improvement of Higher Education Personnel (CAPES) (DS-Finance Code 001), Center for Energy Planning (NIPE), PRP/UNICAMP (PIBIC grant 2327/20), and Technological Development and Innovative Extension Grants (CNPq) (productivity grant 302451/2021-8). R.A. Lamparelli thanks the

National Council for Scientific and Technological Development (CNPq) (research grant 305271/2020-2).

References

Ajao O., Benali M., Faye A., Li H., Maillard D., Ton-That M. T. (2021). Multi-Product Biorefinery System for Wood-Barks Valorization into Tannins Extracts, Lignin-Based Polyurethane Foam and Cellulose-Based Composites: Techno-Economic Evaluation, *Industrial Crops and Products*, 113435.

AliAkbari R., Ghasemi M. H., Neekzad N., Kowsari E., Ramakrishna S., Mehrali M., Marfavi Y. (2021). High Value Add Bio-Based Low-Carbon Materials: Conversion Processes and Circular Economy, *Journal of Cleaner Production*, 293, 126101.

Andrade A., Herrera O. E., Reyes-Contreras P., Pereira M., Vásquez-Garay F. (2021). Eucalyptus Globulus Bark Valorization: Production of Fibers by Neutral Sulphite Semi-Chemical Process for Liner Paper Manufacture, *Floresta e Ambiente*, 28, e20200048.

Clauser N. M., Felissia F. E., Area M. C., Vallejos M. E. (2021). A Framework for the Design and Analysis of Integrated Multi-Product Biorefineries from Agricultural and Forestry Wastes, *Renewable and Sustainable Energy Reviews*, 139, 110687.

Dababi I., Gimello O., Elaloui E., Brosse N. (2020). Water Extraction of Tannins from Aleppo Pine Bark and Sumac Root for the Production of Green Wood Adhesives, *Molecules* 25, 5041.

Dornelles L. B., Filho R. M., Mariano A. P. (2021). Organosolv Fractionation of Eucalyptus: Economics of Cellulosic Ethanol and Chemicals versus Lignin Valorization to Phenols and Polyols, *Industrial Crops and Products*, 173, 114097.

Foelkel C. (Ed.). (2016). *Use of Eucalyptus Biomass for Heat, Steam and Electricity Production*, Eucalyptus Online Book & Newsletter, Brazil.

Junginger H. M., Mai-Moulin T., Daioglou V., Fritsche U., Guisson R., Hennig C., Thrän D., Heinimö J., Hess J. R., Lamers P., Li C., Kwant K., Olsson O., Proskurina S., Ranta T., Schipfer F., Wild M. (2019). The Future of Biomass and Bioenergy Deployment and Trade: A Synthesis of 15 Years IEA Bioenergy Task 40 on Sustainable Bioenergy Trade, Biofuels, *Bioproducts and Biorefining*, 13, 247–266.

Le Maire G., Dupuy S., Nouvellon Y., Loos R. A., Hakamada R. (2014). Mapping Short Rotation Plantations at Regional Scale Using MODIS Time Series: Case of Eucalypt Plantations in Brazil, *Remote Sensing of Environment*, 152, 136–149.

Luhas J., Mikkilä M., Kylkilahti E., Miettinen J., Malkamäki A., Pätäri S., Korhonen J., Pekkanen T. L., Tuppura A., Lähtinen K., Autio M., Linnanen L., Ollikainen M., Toppinen A. (2021). Pathways to a Forest-Based Bioeconomy in 2060 within Policy Targets on Climate Change Mitigation and Biodiversity Protection, *Forest Policy and Economics*, 131, 102551.

Millati R., Cahyono R. B., Ariyanto T., Nafi I., Stp A., Utami R., Taherzadeh M. J. (2019). Chapter 1 - Agricultural, Industrial, Municipal, and Forest Wastes: An Overview,

Chapter In Taherzadeh M. J. (Ed.), *Sustainable Resource Recovery and Zero Waste Approaches*, Vol 1, Elsevier B.V., pp. 1-22.

Orejuela-Escobar L. M., Landázuri A. C., Goodell B. (2021). Second Generation Biorefining in Ecuador: Circular Bioeconomy, Zero Waste Technology, Environment and Sustainable Development: The Nexus, *Journal of Bioresources and Bioproducts*, 6, 83–107.

Parajuli R., Dalgaard T., Jørgensen U., Adamsen A. P. S., Knudsen M. T., Birkved M., Gylling M., Schjørring J. K. (2015). Biorefining in the Prevailing Energy and Materials Crisis: A Review of Sustainable Pathways for Biorefinery Value Chains and Sustainability Assessment Methodologies, *Renewable and Sustainable Energy Reviews*, 43, 244–263.

Pincelli A. L. S. M., Moura L. F., Brito J. O. (2017). Quantification of Harvest Residues in Eucalyptus grandis and Pinus caribaea var. hondurensis forests, *Scientia Forestalis*, 45, 519–526.

Pinto P. C. R., Sousa G., Crispim F., Silvestre A. J. D., Neto C. P. (2013). Eucalyptus globulus Bark as Source of Tannin Extracts for Application in Leather industry, *ACS Sustainable Chemistry & Engineering*, 1, 950–955.

Popp J., Lakner Z., Harangi-Rákos M., Fári M. (2014). The Effect of Bioenergy Expansion: Food, Energy, and Environment, *Renewable and Sustainable Energy Reviews*, 32, 559–578.

Romero C. W. S., Berni M. D., Figueiredo G. K. D. A., Franco T. T., Lamparelli R. A. C. (2019). Assessment of Agricultural Biomass Residues to Replace Fossil Fuel and Hydroelectric Power Energy: A Spatial Approach, *Energy Science & Engineering*, 7, 2287–2305.

SIMA, 2020, *Secretariat of Infrastructure and Environment, Renewable Energy Statistics*, State of São Paulo <dadosenergeticos.energia.sp.gov.br> accessed on 09.01.2022.

Stafford W., De Lange W., Nahman A., Chunilall V., Lekha P., Andrew J., Johakimu J., Sithole B., Trotter D. (2020). Forestry Biorefineries, *Renewable and Sustainable Energy Reviews*, 154, 461–475.

UNICA (2020). *Sugarcane Industry Union, Statistics on hydrous ethanol consumption* <unicadata.com.br/historico-de-consumo-de-combustiveis.php> accessed on 09.01. 2022.

Vital B. R., Carneiro A. C. O., Pimenta A. S., Della Lucia R. M. (2004). Two Eucalypts Bark Tannin-Based Adhesive for Production of Flakeboards, *Revitsta Árvore*, 28, 571–582.

Wang C., Zhang X., Liu Q., Zhang Q., Chen L., Ma L. (2020). A Review of Conversion of Lignocellulose Biomass to Liquid Transport Fuels by Integrated Refining Strategies, *Fuel Processing Technology*, 208, 106485.

Bibliography

Advanced catalysis processes in petrochemicals and petroleum refining: emerging research and opportunities

LCCN	2018042437
Type of material	Book
Main title	Advanced catalysis processes in petrochemicals and petroleum refining: emerging research and opportunities / edited by Mohammed C. Al-Kinany and Saud A. Aldrees.
Published/Produced	Hershey PA: Engineering Science Reference, [2019]
Description	257 pages: illustrations; 25 cm
ISBN	9781522580331 (hardcover) (ebook)
LC classification	TP690 .A235 2019
Related names	Al-Kinany, Mohammed C., 1950- editor. Aldrees, Saud A., 1978- editor.
Summary	"This book examines the technology advancements in the field of refining and petrochemical industries. It also demonstrates how to integrate the refining products within the petrochemical industries"-- Provided by publisher.
Contents	"Catalyst In Biorefineries": Solution To Promote Environment Sustainability In India / Vikas Gupta - New Catalytic Approaches for Producing Alternative to MTBE: Additives for Reformulation of Gasoline / Saud Aldrees - Advanced Catalysis and Processes to Convert Heavy Residues into Fuels and High Value Chemicals / Feras Alshehri, Mohammed Al-Kinany, Saeed Al-Shihri, Bandar Al-Hudaib, Abdulaziz Al-Ghashem, Ali Algarni, Sami Alzahrani, Peter Edwards, Tiancun Xiao - Catalysis in Alkylation of Benzene with Ethene and Propene to Produce Ethylbenzene and Isopropylbenzene / Mohammed AlKinany, Saud

Aldrees, Saeed Alshihri, Eyad Alghilan, Sami Aldrees, Khawla Almalahi, Norah Almousa, Faisal Alotaibi, Yousef Al-Rashed, Feras A. Alshehry - Innovative Catalysis in Petroleum Refinery / Senthil Kumar Ponnusamy, Janet Joshiba G, SSN College of Engineering - Direct conversion of crude oil into high-value chemicals / Senthil Kumar Ponnusamy, Grace Pavitha K - Acetone Reaction with Hydrogen over Mesoporous Magnesium Oxide-Supported Rhodium Nanoparticles / Abdulaziz Bagabas, Ahmad Alshammari, Mohamed Mokhtar, Rasheed Alrasheed, Abdulaziz Asayegh - Conversion of Carbon dioxide to high value products / Nibedita Nath - Development of an engineered microalgal strain for biofuel production / Arti Sharma - Integrated membrane system for the direct Conversion of n-Butane to Isobutene / Adele Brunetti, Leonardo Melone, Ilaria Mirabelli, Enrico Drioli, Giuseppe Barbieri

LC Subjects	Petroleum--Refining. Petroleum chemicals. Catalysis.
Notes	Includes bibliographical references.

Advances in energy systems engineering

LCCN	2019749481
Type of material	Book
Main title	Advances in Energy Systems Engineering / edited by Georgios M. Kopanos, Pei Liu, Michael C. Georgiadis.
Edition	1st ed. 2017.
Published/Produced	Cham: Springer International Publishing: Imprint: Springer, 2017.
Description	1 online resource (XI, 839 pages 395 illustrations, 313 illustrations in color.) PDF
ISBN	9783319428031
Related names	Georgiadis, Michael C. editor. Kopanos, Georgios M. editor.

	Liu, Pei. editor.
Summary	This book provides a scientific framework for integrated solutions to complex energy problems. It adopts a holistic, systems-based approach to demonstrate the potential of an energy systems engineering approach to systematically quantify different options at various levels of complexity (technology, plant, energy supply chain, mega-system). Utilizing modeling, simulation and optimization-based frameworks, along with a number of real-life applications, it focuses on advanced energy systems including energy supply chains, integrated biorefineries, energy planning and scheduling approaches and urban energy systems. Featuring contributions from leading researchers in the field, this work is useful for academics, researchers, industry practitioners in energy systems engineering, and all those who are involved in model-based energy systems.
Contents	1 - Design of sustainable energy systems. 2.-Design and optimization of energy supply chains. 3 - Integrated design of biorefineries. 4 - Energy polygeneration systems. 5 - Integrated oil and gas production. 6 - Urban energy systems. 7 - Production of biofuels. 8 - Energy savings and energy efficiency in the process industries. 9 - Energy planning and scheduling. 10 - Optimal integration of renewable energies.
LC Subjects	Biotechnology. Energy systems. Environmental engineering. Renewable energy sources.
Other Subjects	Energy Systems. Environmental Engineering/Biotechnology. Renewable and Green Energy.
Additional formats	Print version: Advances in energy systems engineering. 9783319428024 (DLC) 2016948241 Printed edition: 9783319428024 Printed edition: 9783319428048

Printed edition: 9783319826677

Agricultural waste: threats and technologies for sustainable management

LCCN	2020058324
Type of material	Book
Main title	Agricultural waste: threats and technologies for sustainable management / Rouf Ahmad Bhat, Khalid Rehman Hakeem, Humaira Qadri, Moonisa Aslam Dervash.
Edition	First edition.
Published/Produced	Palm Bay, FL, USA: Apple Academic Press, 2021.
Description	1 online resource
ISBN	9781003105046 (ebook) (hardcover)
LC classification	TD930
Related names	Bhat, Rouf Ahmad, 1981- editor. Hakeem, Khalid Rehman, editor. Qadri, Humaira, editor. Dervash, Moonisa Aslam, editor.
Summary	"This new volume, Agricultural Waste: Threats and Technologies for Sustainable Management, focuses on tackling the harmful impact of agricultural waste with the use of sustainable practices, bio-techniques, and new technologies that either make use of the waste as a raw material or convert the waste into a useful resource that can help achieve environmental sustainability. The agriculture sector is one of the major producers of waste around the world. These wastes are produced at various stages of the various processes in the agricultural industry. Agricultural practices such as mechanical tillage, mono-cropping, application of agro-chemicals, irrigation with waste and industrial waters affect soil health and productivity. The resulting agricultural wastes can have a deleterious impact on the different components of the environment, resulting in harmful problems that pose a huge risk to ecological stability. This

volume investigates the sources and impacts of agri-waste on the environment and goes on to offer effective techniques, processes, and sustainable practices that can be implemented for effective agricultural waste management. Chapters also discuss a selection of innovative engineering-based technologies. The volume includes discussions of many remediation techniques, such as the use of biofertilizers, biocontrol agents, vermicomposting, green chemistry, microbial interventions, and more. The book explores the various uses of agri-waste with special emphasis on their applications in the plant-soil system. This book is a valuable reference source for academicians, professionals, researchers, students, and policymakers who are interested in the innovative management of the agricultural waste"-- Provided by publisher.

Contents Agricultural Waste: Sources, Implications, and Sustainable Management / Dig Vijay Singh, Rouf Ahmad Bhat, and Syed Maqbool Geelani - Agricultural Wastes and Its Applications in Plant-Soil Systems / Zia Ur Rahman Farooqi, Umair Mubarak, Nukshab Zeeshan, Muhammad Mahroz Hussain, and Muhammad Ashar Ayub - Impact of Agriculture on Soil Health / Mansha Nisar - Global Scenario of Remediation Techniques to Combat Pesticide Pollution / Rezwana Assad, Iqra Bashir, Iflah Rafiq, Irshad Ahmad Sofi, Showkat Hamid Mir, Zafar Ahmad Reshi, and Irfan Rashid - Woodchip Bioreactors for Nitrate Removal in Agricultural Land Drainage / Mehraj U. Din Dar, Aamir Ishaq Shah, Syed Rouhullah Ali, and Shakeel Ahmad Bhat - Biocontrol Agents in Organic Agriculture / Ajaz Ahmad Kundoo, Moonisa Aslam Dervash, Rouf Ahmad Bhat, Barkat Hussain, and Muntazir Mushtaq - The Science of Vermicomposting for Sustainable Development / Rohaya Ali and Rumisa Nazir - Consolidation of Green Chemistry into

Biorefineries: A Pavement for Green and Sustainable Products / Nowsheeba Rashid, Amir Hussain Dar, and Ifra Ashraf - Microbial Interventions and Biochemistry Pathways for Degradation of Agricultural Waste / Monica Butnariu, Ramona Stef, and Alina Butu - Values of Biofertilizers / Hilal Ahmad Ganaie, Naseer Ue Din Shah, Falak Mushtaq, and Jasbir Kour - Alternatives to Synthetic Fertilizers / Muhammad Ijaz, Ijaz Hussain, Muhammad Tahir, Muhammad Shahid, Sami Ul-Allah, Mohsin Zafar, Iqra Rasheed, and Ahmad Nawaz - Mushroom Cultivation Technology for Conversion of Agro-Industrial Wastes into Useful Products / Shauket Ahmed Pala, Dig Vijay Singh, Abdul Hamid Wani, Rouf Ahmad Bhat, and Bashir Ahmad Ganai - An Essay on Some Biotechnological Interventions in Agricultural Waste Management / Rukhsana Akhtar, Adil Farooq Wali, Saiema Rasool, Sabhiya Majid, Hilal Ahmad Wani, Muneeb U. Rehman, Showkat Ahmad Bhat, Shabhat Rasool, Shafat Ali, and Rehan Khan - Bioremediation Technologies for the Management of Agricultural Waste / Monica Butnariu, Ioan Sarac, and Alina Butu.

LC Subjects	Agricultural wastes. Sustainable agriculture. Agricultural innovations.
Notes	Includes bibliographical references and index.
Additional formats	Issued also in electronic format. Print version: Agricultural waste First edition. Palm Bay, FL, USA: Apple Academic Press, 2021. 9781771889636 (DLC) 2020058323

Algal biorefineries and the circular bioeconomy

LCCN	2021050108
Type of material	Book
Main title	Algal biorefineries and the circular bioeconomy / edited by Sanjeet Mehariya, Shashi Kant Bhatia, Obulisamy Parthiba Karthikeyan.

Edition	First edition.
Published/Produced	Boca Raton: CRC Press, 2022.
Description	volumes cm
ISBN	9781032048918 (v. 1; hardback) 9781032049724 (v. 1; paperback) 9781032048932 (v. 2; hardback) 9781032049748 (v. 2; paperback) (v. 1; ebook) (v. 2; ebook)
LC classification	TP339 .A437 2022
Related names	Mehariya, Sanjeet, editor. Bhatia, Shashi Kant, editor. Karthikeyan, Obulisamy Parthiba, editor.
Summary	"This book offers complete coverage of microalgae refineries, including biology, production techniques, biotechnological applications, economic perspectives of applications, and environmental effects of microalgae cultivation. It also summarizes the strategies and future perspectives of microalgal refineries with circular bioeconomy concepts"-- Provided by publisher.
Contents	v. 1. Algal products and processes - v. 2. Industrial applications and future prospects.
LC Subjects	Algal biofuels. Biomass energy industries. Biotechnology--Economic aspects.
Notes	Includes bibliographical references and index.
Additional formats	Online version: Algal biorefineries and the circular bioeconomy. First edition Boca Raton: CRC Press, 2022 9781003195405 (DLC) 2021050109

Algal biorefineries and the circular bioeconomy

LCCN	2021050109
Type of material	Book
Main title	Algal biorefineries and the circular bioeconomy / edited by Sanjeet Mehariya, Shashi Kant Bhatia, Obulisamy Parthiba Karthikeyan.
Edition	First edition.
Published/Produced	Boca Raton: CRC Press, 2022.

Description	1 online resource
ISBN	9781003195429 (v. 2; ebook) (v. 1; hardback) (v. 1; paperback) (v. 2; hardback) 9781003195405 (v. 1; ebook) (v. 2; paperback)
LC classification	TP339
Related names	Mehariya, Sanjeet, editor. Bhatia, Shashi Kant, editor. Karthikeyan, Obulisamy Parthiba, editor.
Summary	"This book offers complete coverage of microalgae refineries, including biology, production techniques, biotechnological applications, economic perspectives of applications, and environmental effects of microalgae cultivation. It also summarizes the strategies and future perspectives of microalgal refineries with circular bioeconomy concepts"-- Provided by publisher.
Contents	v. 1. Algal products and processes - v. 2. Industrial applications and future prospects.
LC Subjects	Algal biofuels. Biomass energy industries. Biotechnology--Economic aspects.
Notes	Includes bibliographical references and index.
Additional formats	Print version: Algal biorefineries and the circular bioeconomy First edition. Boca Raton: CRC Press, 2022 9781032048918 (DLC) 2021050108

Algal biorefining:: resource expenditure and exergo-environmental sustainability

LCCN	2022020383
Type of material	Book
Personal name	Ofori-Boateng, Cynthia, author.
Main title	Algal biorefining:: resource expenditure and exergo-environmental sustainability / Cynthia Ofori-Boateng
Published/Produced	Hauppauge, NY: Nova Science Publishers, [2022]
Description	1 online resource

ISBN	9781685079215 (adobe pdf) (hardcover)
LC classification	TP339
Summary	"This book details the attractive features of algae as a sustainable third-generation feedstock for biorefineries. The commonly explored algal biofuels technologies such as pyrolysis, fermentation, transesterification, gasification, and briquetting/pelleting are assessed in a biorefinery concept and discussed based on resource utilization and sustainability considering biorefinery concepts. Process improvement options for sustainable algal biorefineries are also discussed in this book. Finally, the thermo-environmental sustainabilities of algal biorefineries using exergy and life cycle assessment (LCA) tools are reviewed and assessed based on literature data"-- Provided by publisher.
Contents	Chapter 1. - Algae as third generation feedstock for sustainable biorefinery - Chapter 2. - Algae cultivation and resource utilization - Chapter 3. - Non-fuel products from algae and resource utilization - Chapter 4. - Bioenergy production from algae and resource utilization - Chapter 5. - Exergo-environmental sustainability of algal biorefineries.
LC Subjects	Algal biofuels. Marine algae--Biotechnology. Algae.
Notes	Includes bibliographical references and index.
Additional formats	Also available online. Print version: Ofori-Boateng, Cynthia. Algal biorefining: Hauppauge, NY: Nova Science Publishers, [2022] 9781685077068 (DLC) 2022020382
Series	Waste and waste management Waste and waste management series.

Algal biorefining:: resource expenditure and exergo-environmental sustainability

LCCN	2022020382
Type of material	Book
Personal name	Ofori-Boateng, Cynthia, author.
Main title	Algal biorefining:: resource expenditure and exergo-environmental sustainability / Cynthia Ofori-Boateng
Published/Produced	Hauppauge, NY: Nova Science Publishers, [2022]
ISBN	9781685077068 (hardcover) (adobe pdf)
LC classification	TP339 .O32 2022
Summary	"This book details the attractive features of algae as a sustainable third-generation feedstock for biorefineries. The commonly explored algal biofuels technologies such as pyrolysis, fermentation, transesterification, gasification, and briquetting/pelleting are assessed in a biorefinery concept and discussed based on resource utilization and sustainability considering biorefinery concepts. Process improvement options for sustainable algal biorefineries are also discussed in this book. Finally, the thermo-environmental sustainabilities of algal biorefineries using exergy and life cycle assessment (LCA) tools are reviewed and assessed based on literature data"-- Provided by publisher.
Contents	Chapter 1. - Algae as third generation feedstock for sustainable biorefinery - Chapter 2. - Algae cultivation and resource utilization - Chapter 3. - Non-fuel products from algae and resource utilization - Chapter 4. - Bioenergy production from algae and resource utilization - Chapter 5. - Exergo-environmental sustainability of algal biorefineries.
LC Subjects	Algal biofuels. Marine algae--Biotechnology. Algae.
Notes	Includes bibliographical references and index.

Additional formats	Also available online. Online version: Ofori-Boateng, Cynthia. Algal biorefining: Hauppauge, NY: Nova Science Publishers, 2022 9781685079215 (DLC) 2022020383
Series	Waste and waste management Waste and waste management series.

An integrated approach for added-value products from lignocellulosic biorefineries: vanillin, syringaldehyde, polyphenols and polyurethane

LCCN	2018953351
Type of material	Book
Main title	An integrated approach for added-value products from lignocellulosic biorefineries: vanillin, syringaldehyde, polyphenols and polyurethane / Alirio Egidio Rodrigues.
Published/Produced	New York, NY: Springer Science+Business Media, 2018.
ISBN	9783319993126

An integrated approach for added-value products from lignocellulosic biorefineries: vanillin, syringaldehyde, polyphenols and polyurethane

LCCN	2019764015
Type of material	Book
Personal name	Rodrigues, Alírio Egídio, author.
Main title	An Integrated Approach for Added-Value Products from Lignocellulosic Biorefineries: Vanillin, Syringaldehyde, Polyphenols and Polyurethane / by Alírio Egídio Rodrigues, Paula Cristina de Oliveira Rodrigues Pinto, Maria Filomena Barreiro, Carina Andreia Esteves da Costa, Maria Inês Ferreira da Mota, Isabel Fernandes.
Edition	1st ed. 2018.
Published/Produced	Cham: Springer International Publishing: Imprint: Springer, 2018.
Description	1 online resource (XII, 166 pages 40 illustrations, 24 illustrations in color.) PDF
ISBN	9783319993133

Related names Barreiro, Maria Filomena, author.
Esteves da Costa, Carina Andreia, author.
Fernandes, Isabel, author.
Ferreira da Mota, Maria Inês, author.
Pinto, Paula Cristina de Oliveira Rodrigues, author.

Summary This book offers the state of the art on the progress and accomplishments of 25 years of research at the Associate Laboratory LSRE-LCM - Laboratory of Separation and Reaction Engineering - Laboratory of Catalysis and Materials on lignin conversion to value-added products and their downstream separation. The first valorisation pathway presented for lignin is its partial depolymerisation by oxidation for the production of low molecular weight phenolic compounds, such as vanillin and syringaldehyde, and the second one is the lignin application as macromonomer for polyurethane synthesis. In this book, the authors present the integration of these two valorisation pathways as an exclusive vision of LSRE-LCM resulting from hands-on experience on reaction and separation processes: the integrated process for lignin valorisation. In this perspective, the lignin is oxidized to simultaneously produce syringaldehyde and vanillin, and the obtained by-products to produce a polyol for lignin-based polyurethanes, completing the lignin value chain. On the perspective of pulp mill-related biorefineries, a valorisation route for eucalyptus bark is also presented, focusing on LSRE-LCM experience on extraction and separation of bioactive polyphenols, giving some insights about further integration of extracted bark on biorefining operations.

Contents Chapter 1: Chemical pulp mills as biorefineries - Chapter 2: Integrated process for vanillin and syringaldehyde production from kraft lignin - Chapter 3: Polyurethanes from recovered and depolymerized lignins - Chapter 4: Polyphenols

	from bark of Eucalyptus globulus. - Chapter 5: Conclusions and future perspectives - Chapter 6: References.
LC Subjects	Chemical engineering. Forest products. Industrial engineering. Production engineering. Renewable energy sources.
Other Subjects	Industrial Chemistry/Chemical Engineering. Industrial and Production Engineering. Renewable and Green Energy. Wood Science & Technology.
Additional formats	Print version: An integrated approach for added-value products from lignocellulosic biorefineries: vanillin, syringaldehyde, polyphenols and polyurethane 9783319993126 (DLC) 2018953351 Printed edition: 9783030075880 Printed edition: 9783319993126 Printed edition: 9783319993140

Aries community scale biorefineries: optimizing distributed biofuel and bioenergy production systems: final project report

LCCN	2022362289
Type of material	Book
Main title	Aries community scale biorefineries: optimizing distributed biofuel and bioenergy production systems: final project report / prepared for California Energy Commission, Clean Transportation Program; prepared by Biodiesel Industries, Inc. (Biodico Sustainable Biorefineries); Russell T. Teall III, Russell (Trey) Teall IV, primary authors; Biodico, Inc.
Published/Produced	[Sacramento, Calif.]: California Energy Commission, [2021]
Description	xxii, 373 pages: color illustrations; 28 cm
Links	http://bibpurl.oclc.org/web/97150 https://www.energy.ca.gov/sites/default/files/2021-12/CEC-600-2021-050.pdf

Related names	Teall, Russell, III, author. Teall, Russell, IV, author. Biodico (Firm), author. Clean Transportation Program (Calif.), issuing body.
LC Subjects	Algal biofuels--Research--California. Algae--Industrial applications. Artemia.
Other Subjects	Artemia. California.
Notes	Cover title. "December 2021." Energy Commission pub. no.: CEC-600-2021-050. Includes bibliographical references (pages 371-373). CEC contract number ARV-10-024

Bio-based solvents

LCCN	2017007717
Type of material	Book
Main title	Bio-based solvents / edited by François Jerome, National Higher Engineering School of Poitiers (ENSIP), University of Poitiers, France, Rafael Luque, Department of Quimica Organica, University of Cordoba, Spain.
Published/Produced	Hoboken, NJ: John Wiley & Sons, Inc., 2017.
ISBN	9781119065395 (cloth)
LC classification	TP247.5 .B56 2017
Variant title	Biobased solvents
Related names	Jerome, François, 1974- editor. Luque, Rafael, editor.
Contents	Glycerol as eco-efficient solvent for organic transformations - Aromatic bio-based solvents (alkylphenols, etc.) - Solvents from wastes (in particular extractibles such as limonene and others) - Deep eutectic and low melting mixtures - Organic carbonates: promising reactive solvents for biorefineries and biotechnology - Life cycle assessments for green solvents - Alkylphenols as

	bio-based solvents: properties, manufacture, and applications.
LC Subjects	Solvents. Green chemistry.
Notes	Includes bibliographical references and index.
Additional formats	Online version: Bio-based solvents Hoboken, NJ: John Wiley & Sons, Inc., 2017 9781119065432 (DLC) 2017008295
Series	Wiley series in renewable resources
Bio-based solvents	
LCCN	2017008295
Type of material	Book
Main title	Bio-based solvents / edited by François Jerome, National Higher Engineering School of Poitiers (ENSIP), University of Poitiers, France, Rafael Luque, Department of Quimica Organica, University of Cordoba, Spain.
Published/Produced	Hoboken, NJ: John Wiley & Sons, Inc., 2017.
Description	1 online resource.
ISBN	9781119065432 (pdf) 9781119065449 (epub)
LC classification	TP247.5
Variant title	Biobased solvents
Related names	Jerome, François, 1974- editor. Luque, Rafael, editor.
Contents	Glycerol as eco-efficient solvent for organic transformations - Aromatic bio-based solvents (alkylphenols, etc.) - Solvents from wastes (in particular extractibles such as limonene and others) - Deep eutectic and low melting mixtures - Organic carbonates: promising reactive solvents for biorefineries and biotechnology - Life cycle assessments for green solvents - Alkylphenols as bio-based solvents: properties, manufacture, and applications.
LC Subjects	Solvents. Green chemistry.
Notes	Includes bibliographical references and index.

Additional formats	Print version: Bio-based solvents Hoboken, NJ: John Wiley & Sons, Inc., 2017 9781119065395 (DLC) 2017007717
Series	Wiley series in renewable resources

Biodegradable waste management in the circular economy: challenges and opportunities

LCCN	2021050905
Type of material	Book
Main title	Biodegradable waste management in the circular economy: challenges and opportunities / edited by Malgorzata Kacprzak Czestochowa University of Technology, Czestochowa, Poland, Eleonore Attard, University of Pau, Pau, France, Kari-Anne Lyng, Kråkerøy, Norway, Helena Raclavska, VSB Technical University of Ostrava, Ostrava, Czech Republic, BalRam Singh, Norwegian University of Life Sciences, As, Norway, Eyob Tesfamariam, University of Pretoria, Pretoria, South Africa, Franck Vandenbulcke, Université de Lille, Lille, France.
Edition	First edition.
Published/Produced	Hoboken, NJ: John Wiley & Sons, Inc., 2022.
Description	1 online resource
ISBN	9781119679851 (epub) 9781119679813 (adobe pdf) (hardback)
LC classification	TD804
Related names	Kacprzak, Małgorzata, editor.
Summary	"This book presents the latest developments in the dynamic field of biodegradable waste management, a critical aspect of the circular economy. The book is divided into six sections covering the following topics: The concept of circular economy: The importance of the circular economy, cradle to cradle marketplace, new products from old wastes Biodegradable waste valorization: Definitions and procedures for the characterization of biodegradable waste,

	biodegradable waste streams (agriculture, industry, municipal), new sources of biodegradable waste, limitations for the use of biodegradable waste in soil, waste collection systems, legal/regulation aspects. Technologies for the production of biodegradable waste products (end-products): Biological methods and hybrid technologies, anaerobic digestion, composting, biorefineries Positive and negative effect of end-products on soil/environment: Chemical contaminants occurrence in end-products, effect of end-products on the soil microbial communities, the use of end-products on soil Life cycle perspective: Nutrients recovery, energy and bio-methane production Challenges and opportunities: The governance and social aspects, biofuels production, matter recovery from biodegradable waste streams - fertilizers, metal recovery, biosorbents and biochar production"-- Provided by publisher.
Partial contents	Circular bioeconomy and sustainability / Işıl Aksan Kurnaz, Elif Damla Arison, and M. Levent Kurnaz - Cradle to cradle marketplace / Małgorzata J. Kacpraz - Valorization to energy / Ewa Neczaj and Anna Grosser.
LC Subjects	Organic wastes--Recycling. Waste management. Compost.
Notes	Includes bibliographical references and index.
Additional formats	Print version: Biodegradable waste management in the circular economy First edition. Hoboken, NJ: John Wiley & Sons, Inc., 2022 9781119679844 (DLC) 2021050904

Biodegradable waste management in the circular economy: challenges and opportunities

LCCN	2021050904
Type of material	Book
Main title	Biodegradable waste management in the circular economy: challenges and opportunities / edited by

	Malgorzata Kacprzak Czestochowa University of Technology, Czestochowa, Poland, Eleonore Attard, University of Pau, Pau, France, Kari-Anne Lyng, Kråkerøy, Norway, Helena Raclavska, VSB Technical University of Ostrava, Ostrava, Czech Republic, BalRam Singh, Norwegian University of Life Sciences, As, Norway, Eyob Tesfamariam, University of Pretoria, Pretoria, South Africa, Franck Vandenbulcke, Université de Lille, Lille, France.
Edition	First edition.
Published/Produced	Hoboken, NJ: John Wiley & Sons, Inc., 2022.
ISBN	9781119679844 (hardback) (adobe pdf) (epub)
LC classification	TD804 .B54 2022
Related names	Kacprzak, Małgorzata, editor.
Summary	"This book presents the latest developments in the dynamic field of biodegradable waste management, a critical aspect of the circular economy. The book is divided into six sections covering the following topics: The concept of circular economy: The importance of the circular economy, cradle to cradle marketplace, new products from old wastes Biodegradable waste valorization: Definitions and procedures for the characterization of biodegradable waste, biodegradable waste streams (agriculture, industry, municipal), new sources of biodegradable waste, limitations for the use of biodegradable waste in soil, waste collection systems, legal/regulation aspects. Technologies for the production of biodegradable waste products (end-products): Biological methods and hybrid technologies, anaerobic digestion, composting, biorefineries Positive and negative effect of end-products on soil/environment: Chemical contaminants occurrence in end-products, effect of end-products on the soil microbial communities, the use of end-

	products on soil Life cycle perspective: Nutrients recovery, energy and bio-methane production Challenges and opportunities: The governance and social aspects, biofuels production, matter recovery from biodegradable waste streams - fertilizers, metal recovery, biosorbents and biochar production"-- Provided by publisher.
Partial contents	Circular bioeconomy and sustainability / Işıl Aksan Kurnaz, Elif Damla Arison, and M. Levent Kurnaz - Cradle to cradle marketplace / Małgorzata J. Kacpraz - Valorization to energy / Ewa Neczaj and Anna Grosser.
LC Subjects	Organic wastes--Recycling. Waste management. Compost.
Notes	Includes bibliographical references and index.
Additional formats	Online version: Biodegradable waste management in the circular economy. First edition. Hoboken: Wiley, 2022 9781119679813 (DLC) 2021050905

Biomass and green chemistry: building a renewable pathway

LCCN	2019760335
Type of material	Book
Main title	Biomass and Green Chemistry: Building a Renewable Pathway / edited by Sílvio Vaz Jr.
Edition	1st ed. 2018.
Published/Produced	Cham: Springer International Publishing: Imprint: Springer, 2018.
Description	1 online resource (X, 252 pages 84 illustrations, 47 illustrations in color.) PDF
ISBN	9783319667362
Related names	Vaz Jr., Sílvio. editor.
Summary	This book investigates the main vegetable biomass types, their chemical characteristics and their potential to replace oil as raw material for the chemical industry, according to the principles of green chemistry. Authors from different scientific and technical backgrounds, from industry and

academia, give an overview of the state of the art and ongoing developments. Aspects including bioeconomy, biorefineries, renewable chemistry and sustainability are also considered, given their relevance in this context. Furthermore, the book reviews green chemistry principles and their relation to biomass, while also exploring the main processes for converting biomass into bioproducts. The need to develop renewable feedstock for the chemical industry to replace oil has been identified as a major strategic challenge for the 21st century. In this context, the use of different types of vegetable biomass - starch, lignocellulosic, oleaginous, saccharide and algae - can be seen as a viable alternative to the use of non-renewable, more expensive raw materials. Furthermore, it offers a model for adding economic value to the agro industrial chains such as soybean, sugarcane, corn and forests, among others. This will in turn contribute to the sustainability of a wide range of chemicals, mainly organics and their transformation processes, which are widely used by modern society.

Contents Chapter 1. Biomass and the Green Chemistry Principles - Chapter 2. Saccharide Biomass for Biofuels, Biomaterials and Chemicals - Chapter 3. Oleaginous Biomass for Biofuels, Biomaterials and Chemicals - Chapter 4. Starch Biomass for Biofuels, Biomaterials and Chemicals - Chapter 5. Lignocellulosic Biomass for Energy, Biofuels, Biomaterials, and Chemicals - Chapter 6. Microalgae for Industrial Purposes - Chapter 7. Enzymatic Conversion of First and Second Generation Sugars - Chapter 8. Sustainability of Biomass.

LC Subjects Chemical engineering.
Green chemistry.
Natural resources.
Renewable energy sources.

Other Subjects	Green Chemistry. Industrial Chemistry/Chemical Engineering. Natural Resources. Renewable and Green Energy.
Additional formats	Print version: Biomass and green chemistry 9783319667355 (DLC) 2017959190 Printed edition: 9783319667355 Printed edition: 9783319667379 Printed edition: 9783319883083

Bioprocessing of biofuels

LCCN	2020037720
Type of material	Book
Personal name	Sarangi, Prakash Kumar, author.
Main title	Bioprocessing of biofuels / author by Prakash Kumar Sarangi and Sonil Nanda.
Edition	First edition.
Published/Produced	Boca Raton: CRC Press, 2021.
ISBN	9780367340841 (hardback) (ebook)
LC classification	TP339 .S366 2021
Related names	Nanda, Sonil, author.
Summary	"The major issues relating to environmental sustainability such as a heavy dependency on fossil fuels, increased greenhouse gas emissions, pollution, global warming and climate change have drawn much attention around the globe to seek alternative energy sources that have negligible environmental impacts and societal benefits. There is an immense interest in biofuels research around the world due to their massive potentials in addressing the above-mentioned environmental concerns. Biofuels have the potential to supplement the current and future energy demands by being blended with fossil fuels or replacing them completely as drop-in fuels in automobiles as well as heat and power industries. Waste biomass, primarily lignocellulosic biomass (e.g., agricultural crop residues, forestry biomass and energy crops)

	and microalgae can act as some inexpensive renewable bioresources for the production of biofuels and biochemicals. The prime focus of this book is to shed light on the bioprocessing of biofuels, especially through microbial conversion technologies to recover and transform the inedible polysaccharides into hydrocarbons biofuels and bioenergy. The book offers introductory coverage of the most crucial topics as follows: A systematic overview of the state-of-the-art in the production and utilization of biofuels Categorical bioprospecting of bioresources for biofuel production Biomass pretreatment and enzymatic saccharification Bioconversion of waste biomass and algae to liquid and gaseous biofuels New developments in microbial fuel cell technologies This book is a unification of topics relating to the cutting-edge applications of bioresources and green technologies to reinvigorate the biorefineries in positioning them within a competitive energy market. Persistently fabricated to be instantly applicable, this volume serves as a reference book for undergraduate and graduate students, scientific investigators and research scholars across the globe working in the areas relating to energy and fuels"--Provided by publisher.
LC Subjects	Biomass energy.
Notes	Includes bibliographical references and index.

Biorefineries: concepts, advancements and research

LCCN	2017005215
Type of material	Book
Uniform title	Biorefineries (Torres)
Main title	Biorefineries: concepts, advancements and research / Irene Torres, editor
Published/Produced	New York: Novinka, [2017]
Description	1 online resource.
ISBN	9781536107203 (Ebook)
LC classification	TP339

Related names Torres, Irene, editor.
LC Subjects Biomass--Refining.
Biomass conversion.
Biomass chemicals.
Notes Includes bibliographical references and index.
Additional formats Print version: Biorefineries New York: Novinka, [2017] 9781536107081 (DLC) 2016055972
Series Petroleum science and technology

Biorefineries: concepts, advancements and research
LCCN 2016055972
Type of material Book
Uniform title Biorefineries (Torres)
Main title Biorefineries: concepts, advancements and research / Irene Torres, editor.
Published/Produced New York: Novinka, [2017]
Description viii, 116 pages; 23 cm.
ISBN 9781536107081 (softcover)
LC classification TP339 .B5865 2017
Related names Torres, Irene, editor.
LC Subjects Biomass--Refining.
Biomass conversion.
Biomass chemicals.
Notes Includes bibliographical references and index.
Additional formats Online version: Biorefineries New York: Novinka, [2017] 9781536107203 (DLC) 2017005215
Series Petroleum science and technology

Biorefineries: design and analysis
LCCN 2018018742
Type of material Book
Personal name Cardona Alzate, Carlos Ariel, author.
Main title Biorefineries: design and analysis / Carlos Ariel Cardona Alzate, Jonathan Moncada Botero, Valentina Aristizábal Marulanda.
Published/Produced Boca Raton: CRC Press, 2019.
ISBN 9781138080027 (hardback: alk. paper)
ebook
LC classification TP339 .C369 2019

Related names	Moncada Botero, Jonathan, author. Aristizábal Marulanda, Valentina, author.
Summary	"Book presents a systematic design of biorefineries, giving an overview about relevance and application including framework for biorefineries based on techno-economic, environmental and social aspects. Mass, energy and economic indices are considered to assess the biorefinery sustainability as well"-- Provided by publisher.
Contents	The biorefinery concept - Biorefinery feedstocks, platforms and products - Biorefinery design strategy: from process synthesis to sustainable design - Techno-economic analysis of biorefineries - Environmental assessment of biorefinery systems - Social analysis of biorefineries - Development of biorefinery systems: from biofuel upgrading to multiproduct portfolios - Sustainability assessment of biorefineries based on indices - Key challenges for future development of biorefineries.
LC Subjects	Biomass conversion--Environmental aspects. Biomass conversion--Economic aspects. Biomass conversion--Social aspects. Biomass--Refining.
Notes	Includes bibliographical references and index.

Biorefineries: design and analysis

LCCN	2020692186
Type of material	Book
Personal name	Cardona Alzate, Carlos Ariel, author.
Main title	Biorefineries: design and analysis / Carlos Ariel Cardona Alzate, Jonathan Moncada Botero, Valentina Aristizábal Marulanda.
Published/Produced	Boca Raton, FL: CRC Press, [2019]
Description	1 online resource.
ISBN	9781315114088 (ebook) 9781351626255 (ebook) (hardback; alk. paper)
LC classification	TP339

Related names	Moncada Botero, Jonathan, author. Aristizábal Marulanda, Valentina, author.
Summary	"Book presents a systematic design of biorefineries, giving an overview about relevance and application including framework for biorefineries based on techno-economic, environmental and social aspects. Mass, energy and economic indices are considered to assess the biorefinery sustainability as well"-- Provided by publisher.
Contents	The biorefinery concept - Biorefinery feedstocks, platforms and products - Biorefinery design strategy: from process synthesis to sustainable design - Techno-economic analysis of biorefineries - Environmental assessment of biorefinery systems - Social analysis of biorefineries - Development of biorefinery systems: from biofuel upgrading to multiproduct portfolios - Sustainability assessment of biorefineries based on indices - Key challenges for future development of biorefineries.
LC Subjects	Biomass conversion--Environmental aspects. Biomass conversion--Economic aspects. Biomass conversion--Social aspects. Biomass--Refining.
Notes	Includes bibliographical references and index.
Additional formats	Print version: Biorefineries Boca Raton: CRC Press, [2019] 9781138080027 (hardback: alk. paper) (DLC) 2018018742

Biorefineries: targeting energy, high value products and waste valorisation

LCCN	2019749380
Type of material	Book
Main title	Biorefineries: Targeting Energy, High Value Products and Waste Valorisation / edited by Miriam Rabaçal, Ana F. Ferreira, Carla A. M. Silva, Mário Costa.
Edition	1st ed. 2017.

Published/Produced	Cham: Springer International Publishing: Imprint: Springer, 2017.
Description	1 online resource (XVIII, 294 pages 110 illustrations, 77 illustrations in color.) PDF
ISBN	9783319482880
Related names	Costa, Mário. editor. Ferreira, Ana F. editor. Rabaçal, Miriam. editor. Silva, Carla A. M. editor.
Summary	This book addresses the potential of the transformation of biomass into a wide range of marketable products, and examines the biological, biochemical, physical and thermal processing of biomass into products such as fuels, power, heat, feeds, chemicals and materials. Respective chapters explore various topics including biomass characterization, biomass pre-conditioning and sustainability analysis, aspects that are supplemented by a global overview of their implementation in current pilot bio-refineries. Providing a valuable resource to energy engineers, chemical engineers, biotechnologists and economists, this book will also be of great interest to students and policymakers.
Contents	Introduction - Biomass characterization - Biomass pre-conditioning - Biomass transformation - Sustainability analysis - Existing Pilot-plants - Conclusions.
LC Subjects	Biotechnology. Energy systems. Environmental economics. Environmental engineering. Renewable energy sources.
Other Subjects	Renewable and Green Energy. Biotechnology. Energy Systems. Energy Systems. Environmental Economics.

Environmental Engineering/Biotechnology.
Additional formats Print version: Biorefineries. 9783319482866 (DLC) 2017932082
Printed edition: 9783319482866
Printed edition: 9783319482873
Printed edition: 9783319839059
Series Lecture Notes in Energy, 2195-1284; 57
Lecture Notes in Energy, 2195-1284; 57

Biorefineries
LCCN 2017932082
Type of material Book
Main title Biorefineries.
Published/Produced New York, NY: Springer Berlin Heidelberg, 2017.
ISBN 9783319482866

Biorefineries
LCCN 2018958916
Type of material Book
Personal name Wagemann, Kurt.
Main title Biorefineries / Kurt Wagemann.
Published/Produced New York, NY: Springer Berlin Heidelberg, 2018.
ISBN 9783319971179 (alk. paper)

Biorefineries
LCCN 2019744523
Type of material Book
Main title Biorefineries / edited by Kurt Wagemann, Nils Tippkötter.
Edition 1st ed. 2019.
Published/Produced Cham: Springer International Publishing: Imprint: Springer, 2019.
Description 1 online resource (VI, 549 pages 159 illustrations, 89 illustrations in color.)
PDF
ISBN 9783319971193
Related names Tippkötter, Nils. editor.
Wagemann, Kurt. editor.

Summary	This book offers a comprehensive review on biomass resources, examples of biorefineries and corresponding products. The first part of this book covers topics such as different biorefinery resources from agriculture, wood processing residues and transport logistics of plant biomass. In the second part, expert contributors present biorefinery concepts of different biomass feedstocks, including vegetable-oils, sugarcane, starch, lignocellulose and microalgae. Readers will find here a summary of the syngas utilization and the bio-oil characterization and potential use as an alternative renewable fuel and source for chemical feedstocks. Particular attention is also given to the anaerobic digestion-based and Organosolv biorefineries. The last part of the book examines relevant products and components such as alcohols, hydrocarbons, bioplastics and lignin, and offers a sustainability evaluation of biorefineries.
Contents	From the Contents: Lignocellulose-Biorefinery: Organosolv-Processes - Syngas-Biorefinery: Biomass-Gasification - Syngas-Biorefinery: Syngas-Utilization - Algal-Biorefinery - Products: Lignin - Biogas-Biorefinery: Syngas-Utilization - Biomass Resources: Agriculture.
LC Subjects	Biotechnology. Chemical engineering. Renewable energy sources.
Other Subjects	Biotechnology. Industrial Chemistry/Chemical Engineering. Renewable and Green Energy. Renewable and Green Energy.
Additional formats	Print version: Biorefineries 9783319971179 (DLC) 2018958916 Printed edition: 9783319971179 Printed edition: 9783319971186
Series	Advances in Biochemical Engineering/ Biotechnology, 0724-6145; 166

Advances in Biochemical Engineering/ Biotechnology, 0724-6145; 166

Biorefining of biomass to biofuels: opportunities and perception

LCCN	2019754951
Type of material	Book
Main title	Biorefining of Biomass to Biofuels: Opportunities and Perception / edited by Sachin Kumar, Rajesh K. Sani.
Edition	1st ed. 2018.
Published/Produced	Cham: Springer International Publishing: Imprint: Springer, 2018.
Description	1 online resource (XVII, 372 pages 113 illustrations, 56 illustrations in color.) PDF
ISBN	9783319676784
Related names	Kumar, Sachin, 1981- editor. Sani, Rajesh K. editor.
Summary	This volume focuses on the prospects of the conversion of biomass into biofuels including ethanol, butanol, biogas, biohydrogen, biodiesel, syn-gas and other useful products. Biomass-derived fuels have gained tremendous attention worldwide. However, due to high raw material and processing costs, biofuels produced from lignocelluloses have been found to be more expensive than conventional fuels. Therefore, a concept of biorefining has been introduced, where more than one product or each and every component of biomass may be derived into useful products in a manner of petroleum refinery.
Contents	Biorefining of lignocelluloses: an opportunity for sustainable biofuel production - Biomass, its potential and applications - Biomass gasification and sustainability assessment of biomass utilization - Advances in transformation of lignocellulosic biomass to carbohydrate derived fuel precursors - Biodiesel synthesis: use of activated carbon as support of the catalysts -

	Biorefining of biomass to biofuels: opportunities and perception - Potential role of halophile in crude glycerol based bio-refinery - Bio-Jet fuel - Pretreatment of lignocellulosic biomass towards biofuel production - Operational strategies for enzymatic hydrolysis in a biorefinery - Prospects of solvent tolerance in butanol fermenting bacteria - Simultaneous saccharification and fermentation of lignocellulosic biomass - Bioalkanes and bioalkenes: an ecofriendly and alternate fuel in bioenergy research - Algal biorefineries for biofuels and other value-added products - Biodiesel -technical viability for India - Kinetic modeling of ethanol production for substrate-microbe system.
LC Subjects	Biochemical engineering. Microbiology. Renewable energy sources.
Other Subjects	Renewable and Green Energy. Biochemical Engineering. Microbiology. Renewable and Green Energy.
Additional formats	Print version: Biorefining of biomass to biofuels. 9783319676777 (DLC) 2017956319 Printed edition: 9783319676777 Printed edition: 9783319676791 Printed edition: 9783319884660
Series	Biofuel and Biorefinery Technologies, 2363-7609; 4 Biofuel and Biorefinery Technologies, 2363-7609; 4

Bioremediation and sustainable technologies for cleaner environment

LCCN	2019752702
Type of material	Book
Main title	Bioremediation and Sustainable Technologies for Cleaner Environment / edited by Marimuthu Prashanthi, Rajakumar Sundaram, Aravind Jeyaseelan, Thamaraiselvi Kaliannan.

Edition	1st ed. 2017.
Published/Produced	Cham: Springer International Publishing: Imprint: Springer, 2017.
Description	1 online resource (XI, 358 pages 136 illustrations, 92 illustrations in color.) PDF
ISBN	9783319484396
Related names	Jeyaseelan, Aravind. editor. Kaliannan, Thamaraiselvi. editor. Prashanthi, Marimuthu. editor. Sundaram, Rajakumar. editor.
Summary	This book offers insights into the current focus and recent advances in bioremediation and green technology applications for waste minimization and pollution control. Increasing urbanization has an impact on the environment, agriculture and industry, exacerbating the pollution problem and creating an urgent need for sustainable and green eco-friendly remediation technology. Currently, there is heightened interest in environmental research, especially in the area of pollution remediation and waste conversion, and alternative, eco-friendly methods involving better usage of agricultural residues as economically viable substrates for environmental cleanup are still required. The book offers researchers and scholars inspiration, and suggests directions for specific waste management and pollution control. The research presented makes a valuable contribution toward a sustainable and eco-friendly societal environment.
Contents	Biological perspective and role of bacteria in pesticide degradation - Nanobioremediation technologies for sustainable environment - Stabilization of market vegetable waste through the process of vermicomposting by *Eisenia foetida* - Nitrate removal from ground water through lab scale bioreactor using dissimilatory nitrate reducer *Bacillus weihenstephanensis* (DS45) - Catalytic

degradation of reactive red 120 by copper oxide nanoparticles synthesized by *Azadirachda indica* - Removal of Nickel from synthetic waste water Using Gooseberry Seeds as biosorbent - Bioconversion Of Cellulosic Waste Into Bioethanol - A Synergistic Interaction Of *Trichoderma viride* and *Saccharomyces cerevisiae* - Efficient hydrolysis of Lignocellulosic Biomass: Potential Challenges and Future Perspectives for Biorefineries - Optimization of media components for production of polyhydroxyalkanoates (bioplastic) by *Ralstonia eutropha* using paddy straw as cheap substrate - Microbial Bioremediation of hazardous heavy metals - Screening and isolation of textile reactive dyes decolorizing indigenous fungi isolated from textile effluent contaminated site - Optimization of biosurfactant production and crude oil emulsification by *Bacillus* sp. isolated from hydrocarbon contaminated soil sample.

LC Subjects Biotechnology.
Environmental engineering.
Pollution.
Renewable energy sources.
Sustainable development.
Waste management.

Other Subjects Waste Management/Waste Technology.
Environmental Engineering/Biotechnology.
Pollution, general.
Renewable and Green Energy.
Sustainable Development.

Additional formats Print version: Bioremediation and sustainable technologies for cleaner environment. 9783319484389 (DLC) 2016960281
Printed edition: 9783319484389
Printed edition: 9783319484402
Printed edition: 9783319839431

Series Environmental Science, 1431-6250

Biosynthetic technology and environmental challenges

LCCN	2019764256
Type of material	Book
Main title	Biosynthetic Technology and Environmental Challenges / edited by Sunita J. Varjani, Binod Parameswaran, Sunil Kumar, Sunil K. Khare.
Edition	1st ed. 2018.
Published/Produced	Singapore: Springer Singapore: Imprint: Springer, 2018.
Description	1 online resource (XV, 401 pages 53 illustrations, 38 illustrations in color.) PDF
ISBN	9789811074349
Related names	Khare, Sunil K, editor. Kumar, Sunil, editor. Binod, Parameswaran, editor. Varjani, Sunita J, editor.
Summary	This book provides a comprehensive review of biosynthetic approaches to the production of industrially important chemicals and the environmental challenges involved. Its 19 chapters discuss different aspects of biosynthetic technology from the perspective of leading experts in the field. It covers various biorefinery approaches, including the use of microbes, metabolically engineered plants, biomass-based and green technology methods. Further, it examines important research in the areas of organic and hazardous waste composting, management and recovery of nutraceuticals from agro-industrial waste, biosynthesis and technological advancements of biosurfactants and waste water bioremediation. This book contributes to the scientific literature on biosynthetic technologies and the related environmental challenges for researchers and academics working in this area around the globe.
Contents	Part I General - Introduction to Biosynthetic Technology and Environmental - Part II:

Biosynthetic Approaches and Products - Management of Agro-Industrial Wastes with the Aid of Synthetic Biology - Advances and Tools in Engineering Yeast for Pharmaceutical Production - Plant Biosynthetic Engineering through Transcription Regulation: An insight into Molecular Mechanisms During Environmental Stress - Oil Palm Biomass and its Kinetic Transformation Properties - Selection and Utilization of Agro-Industrial Waste for Biosynthesis and Hyper-Production of Pullulan: A Review - Production, Characterization, and Applications of Microbial Poly-Glutamic Acid - Bioprocesses for the Production of 2,5-Furandicarboxylic Acid - Biosynthesis of 1,3-Propanediol: Genetics and Applications - Biosynthesis and Technological Advancements of Biosurfactants - Recovery of Nutraceuticals from Agri-Food Industry Waste by Lactic Acid Fermentation - Manno-Oligosaccharides as Prebiotic Valued Products from Agro-Waste - Computational Modelling and Prediction of Microalgae Growth Focused towards Improved Lipid Production - Perennial Energy Crops on Drained Peatlands in Finland - Part III: Environmental Assessment and Waste Management - Bioenergy Conversion from Aquatic Weed Water Hyacinth into Agronomically Valuable Vermicompos - Mitigation of Global Warming Potential for Cleaner Composting - Recent Advances in Composting of Organic and Hazardous Waste: A Road Map to Safer Environment - Biomarkers Currently used in Water Pollution Monitoring Program - Bioremediation by Microalgae: Current and Emerging Trends for Effluents Treatments for Value Addition of Waste Streams - Environmental Assessment of Biorefineries.

LC Subjects Biomedical materials.

	Biomedical engineering.
	Biotechnology.
	Bioengineering--Materials.
	Waste management.
Other Subjects	Biomaterials.
	Biomedical Engineering/Biotechnology.
	Biotechnology.
	Waste Management/Waste Technology.
Additional formats	Print version: Biosynthetic technology and environmental challenges. 9789811074332 (DLC) 2017959152
	Printed edition: 9789811074332
	Printed edition: 9789811074356
	Printed edition: 9789811356339
Series	Energy, Environment, and Sustainability, 2522-8366
	Energy, Environment, and Sustainability, 2522-8366

Biotechnology of yeasts and filamentous fungi

LCCN	2019748524
Type of material	Book
Main title	Biotechnology of Yeasts and Filamentous Fungi / edited by Andriy A. Sibirny.
Edition	1st ed. 2017.
Published/Produced	Cham: Springer International Publishing: Imprint: Springer, 2017.
Description	1 online resource (VIII, 412 pages 48 illustrations, 21 illustrations in color.)
	PDF
ISBN	9783319588292
Related names	Sibirny, Andriy A. editor.
Summary	This book provides a comprehensive overview on biotechnological applications of unicellular and multicellular fungi in a variety of industrial branches. Targeted genetic and metabolic engineering of fungi allows production of native and transgenic enzymes and proteins in industrial scales. Those most prominently find application in

	biorefineries for the production of value-added chemicals and biofuels, in the pharmaceutical industry as well as in biomedicine. Each chapter is dedicated to applications and potential beneficial use of particular strains of yeasts and filamentous fungi and their produced biomolecules. The book targets researchers from both academia and industry and graduate students working in microbial biotechnology.
Contents	1.Genetic improvement of conventional and non-conventional yeasts for the production of 1st and 2nd generation ethanol - 2.High-temperature bio-ethanol fermentation by conventional and non-conventional yeasts - 3.Construction of baker's yeast strains with enhanced tolerance to baking-associated stresses - 4.Anhydrobiosis and dehydration of yeasts - 5.Biotechnology of glycerol production and conversion in yeasts - 6.Lipids of yeasts and filamentous fungi and their importance for biotechnology - 7.Production of Organic Acids by Yeasts and Filamentous Fungi - 8.Carotenoid production by filamentous fungi and yeasts - 9.Molecular studies of the flavinogenic fungus *Ashbya gosspypii* and the flavinogenic yeast Candida famata - 10.Cold-active enzymes from cold-adapted yeasts - 11.Methylotrophic yeasts as producers of recombinant proteins - 12.Biosensors based on yeast/fungal cells - 13.East-Based Systems For Environmental Control - 14.Yeast-Based Biosensors For Clinical Diagnostics And Food Control.
LC Subjects	Microbial genetics. Microbial genomics. Microbiology. Mycology.
Other Subjects	Microbial Genetics and Genomics. Applied Microbiology. Mycology.

Additional formats	Print version: Biotechnology of yeasts and filamentous fungi 9783319588285 (DLC) 2017943864 Printed edition: 9783319588285 Printed edition: 9783319588308 Printed edition: 9783319864853

Circular bioeconomy - current developments and future outlook

LCCN	2021936274
Type of material	Book
Main title	Circular bioeconomy - current developments and future outlook / Ashok Pandey, R. D. Tyagi, Sunita Varjani.
Edition	1.
Published/Produced	Cambridge: Elsevier, 2021.
ISBN	9780128218785 (paperback)
Related names	Pandey, Ashok, editor. Tyagi, R. D., editor. Varjani, Sunita, editor.
Summary	"Biomass, Biofuels, Biochemical: Circular Bioeconomy: Current Developments and Future Outlook presents the views of experienced academicians and researchers working in the area of the circular bioeconomy. This book presents an assortment of Resource recovery, Waste Biorefineries, Bio-electrochemical systems, Biopolymers and Green solvents, Bio-adsorbents, and Technology transfer topics. Environmental engineers, biotechnologists, science graduates, chemical engineers, industrial experts and policymakers working in these areas will find the information on the circular economy and its important part in developing low carbon and resource-productive economies very informative. Methodologies and beneficial strategic approaches to address the usage of wastes from agriculture, co-products, and by-products are also discussed"--Provided by publisher.

Economics of bioresources: concepts, tools, experiences

LCCN	2019738016
Type of material	Book
Main title	Economics of Bioresources: Concepts, Tools, Experiences / edited by Yoram Krozer, Michael Narodoslawsky.
Edition	1st ed. 2019.
Published/Produced	Cham: Springer International Publishing: Imprint: Springer, 2019.
Description	1 online resource (XI, 210 pages 17 illustrations, 10 illustrations in color.) PDF
ISBN	9783030146184
Related names	Krozer, Yoram. editor. Narodoslawsky, Michael. editor.
Summary	This fundamental book provides a cross-sectoral, multi-disciplinary view on the biobased economy. It explains opportunities for the value-adding production and use of bioresources, while also discussing the main drivers and obstacles involved. The book is divided into three major parts, the first of which introduces readers to the basics of bioresource economics and engineering. In terms of economics, it discusses decision-making from the policy, producer, investor, and citizen perspectives; in terms of engineering, it addresses key technologies and the processing of bioresources, as well as the development of biorefineries for high-value products on large and small scales. In turn, the book's second part presents cases focused on different types of energy use, and written by practitioners. The cases illustrate the businesses and technologies involved, as well as the roles of citizens, social organisations and policies. The book's third and last part highlights opportunities in sustainable agriculture, valuable industrial products and innovative services, while also outlining key conditions for success. Written by a team of scholars and

	practitioners from various engineering, natural-science and social-science disciplines, the book is primarily intended for undergraduate and graduate students, and for practitioners in business and policy who wish to explore the sustainable production and use of bioresources. All theoretical issues are explained with the aid of real-world examples, making the content highly accessible.
Contents	1.Introduction: Yoram Krozer, Maarten Arentsen, Michael Narodoslawsky (10 pages) - Part I: Basics (20-25 pages each chapter) - 2.Bioresources basics: Michael Narodoslawsky - 3.Bioresources for electricity: Andrej Gubina - 4.Economic basics: Yoram Krozer - 5.Regional capabilities: Frans Coenen - 6.Market behavior: Boglarka Vajda and Yoram Krozer - 7.Business development: Yoram Krozer and Petra Doelman - 8.Policy development: Maarten Arentsen - Part II Cases (5-10 pages each chapter) - 9.Willows for localized fuels: Sandor Bartha - 10.Modern bio-refineries: Louis Duarte - 11.Supplies for bioenergy markets: Devrim Yazan - 12.Food chain waste processing: Ina Körner - 13.Reuse of edible oils: Avinash Narayanswamy - 14.Manure for energy: Willem van Laarhoven.
LC Subjects	Natural resources. Renewable energy sources. Energy policy. Energy and state. Chemical engineering. Electrical engineering.
Other Subjects	Natural Resource and Energy Economics. Renewable and Green Energy. Energy Policy, Economics and Management. Industrial Chemistry/Chemical Engineering. Electrical Engineering.
Additional formats	Printed edition: 9783030146177 Printed edition: 9783030146191 Printed edition: 9783030146207

Efficiency of biomass energy: an exergy approach to biofuels, power, and biorefineries

LCCN	2016024267
Type of material	Book
Personal name	Ptasinski, Krzysztof J., 1946- author.
Main title	Efficiency of biomass energy: an exergy approach to biofuels, power, and biorefineries / Krzysztof J. Ptasinski.
Published/Produced	Hoboken, New Jersey: Wiley, [2016]
Description	1 online resource.
ISBN	9781119118145 (epub) 9781119118152 (pdf)
LC classification	TP339
Summary	"Details energy and exergy efficiencies of all major aspects of bioenergy systems Covers all major bioenergy processes starting from photosynthesis and cultivation of biomass feedstocks and ending with final bioenergy products, like power, biofuels, and chemicals Each chapter includes historical developments, chemistry, major technologies, applications as well as energy, environmental and economic aspects in order to serve as an introduction to biomass and bioenergy A separate chapter introduces a beginner in easy accessible way to exergy analysis and the similarities and differences between energy and exergy efficiencies are underlined Includes case studies and illustrative examples of 1st, 2nd, and 3rd generation biofuels production, power and heat generation (thermal plants, fuel cells, boilers), and biorefineries; Traditional fossil fuels-based technologies are also described in order to compare with the corresponding bioenergy systems."-- Provided by publisher.
Contents	Machine generated contents note: Preface Acknowledgments About the Author Part I Background and Outline 1. Bioenergy Systems - An Overview 1.1 Energy and the Environment 1.2 Biomass as a Renewable Energy Source 1.3

Biomass Conversion Processes 1.4 Utilization of Biomass 1.5 Closing Remarks References 2. Exergy Analysis 2.1 Sustainability and Efficiency 2.2 Thermodynamic Analysis of Processes 2.3 Exergy Concept 2.4 Exergetic Evaluation of Processes and Technologies 2.5 Renewability of Biofuels 2.6 Closing Remarks References Part II Biomass Production and Conversion 3. Photosynthesis 3.1 Photosynthesis - an Overview 3.2 Exergy of Thermal Radiation 3.3 Exergy Analysis of Photosynthesis 3.4 Global Photosynthesis 3.5 Closing Remarks References 4. Biomass Production 4.1 Overview 4.2 Efficiency of Solar Energy Capture 4.3 Fossil Inputs for Biomass Cultivation and Harvesting 4.4. Fossil Inputs for Biomass Logistics 4.5 Closing Remarks References 5 Thermochemical Conversion: Gasification 5.1 Gasification - an Overview 5.2 Gasification of Carbon 5.3 Gasification of Biomass 5.4 Gasification of Typical Fuels 5.5 Closing Remarks References 6 Thermochemical Conversion: Parametric Studies And Gasification Systems 6.1 Effect of Fuel Chemical Composition on Gasification Performance 6.2 Effect of Biomass Moisture Content, Gasification Pressure and Heat Addition on Gasification Performance 6.3 Improvement of Gasification Exergetic Efficiency 6.4 Gasification Efficiency Using Equilibrium versus Non-equilibrium Models 6.5 Performance of Typical Gasifiers 6.6 Plasma Gasification 6.7 Thermochemical Conversion in Sub- and Supercritical Water 6.8 Closing Remarks References Part III Biofuels First-Generation Biofuels 7 Biodiesel 7.1 Biodiesel: an Overview 7.2 Biodiesel from Plant Oils 7.3 Biodiesel from Waste Cooking Oil 7.4 Biodiesel from Microalgae 7.5 Closing Remarks References 8 Bioethanol 8.1 Bioethanol - an Overview 8.2 Exergy Analysis of Ethanol from Sugar Crops 8.3 Exergy Analysis of

Ethanol from Starchy Crops 8.4 Exergy Analysis of Lignocellulosic Ethanol (2nd Generation) 8.5 Alternative Ethanol Processes 8.6 Closing Remarks References Second-Generation Liquid Biofuels 9 Fischer-Tropsch Fuels 9.1 Fischer-Tropsch Synthesis - an Overview 9.2 Exergy Analysis of Coal-to-Liquid (CTL) Process 9.3 Exergy Analysis of Gas-to-Liquid (GTL) Processes 9.4 Exergy Analysis of Biomass-to-Liquid (BTL) Processes 9.5 Closing Remarks References 10 Methanol 10.1 Methanol - an Overview 10.2 Methanol from Fossil Fuels 10.3 Methanol from Biomass 10.4 Closing Remarks References 11 Thermochemical Ethanol 11.1 Thermochemical Ethanol -an Overview 11.2 Exergy Analysis 11.3 Closing Remarks References Second-Generation Gaseous Biofuels 12 Dimethyl Ether (DME) 12.1 Dimethyl Ether - an Overview 12.2 Dimethyl Ether from Fossil Fuels 12.3 Dimethyl Ether from Biomass 12.4 Closing Remarks References 13 Hydrogen 13.1 Hydrogen - an Overview 13.2 Exergy Analysis of Hydrogen from Fossil Fuels 13.3 Exergy Analysis of Hydrogen from Water Electrolysis 13.4 Exergy Analysis of Future Hydrogen Production Processes 13.5 Exergy Analysis of Hydrogen from Biomass Gasification 13.6 Exergy Analysis of Biological Hydrogen Production 13.7 Closing Remarks References 14 Substitute Natural Gas (SNG) 14.1 Substitute Natural Gas - an Overview 14.2 SNG from Coal 14.3 SNG from Biomass Gasification 14.4 Closing Remarks References Part IV Bioenergy Systems 15 Thermal Power Plants, Heat Engines And Heat Production 15.1 Biomass-Based Power and Heat Generation - an Overview 15.2 Biomass Combustion Power Systems 15.3 Biomass Gasification Power Systems 15.4 Comparison of Various Biomass-Fueled Power Plants 15.5 Biomass-Fueled Internal Combustion

	Engines and Gas Turbines 15.6 Polygeneration of Electricity, Heat and Chemicals 15.7 Biomass Boilers and Heating Systems 15.8 Closing Remarks References 16 Biomass-Based Fuel Cell Systems 16.1 Biomass-Based Fuel Cell Systems - an Overview 16.2 Biomass Integrated Gasification: Solid Oxide Fuel Cell (BIG/SOFC) Systems 16.3 Biomass Integrated Gasification: Proton Exchange Membrane Fuel Cell (BIG/PEMFC) Systems 16.4 Fuel Cell Systems Fed with Liquid Biofuels 16.5 Closing Remarks References 17 Biorefineries 17.1 Biorefineries: an Overview 17.2 Comparison of Various Biomass Utilization Routes 17.3 Exergy Inputs to Basic Biorefinery Steps 17.4 Optimal Biomass Crops as Biorefinery Feedstock 17.5 Closing Remarks References Postface Appendixes Appendix A: Conversion Factors Appendix B: Constants Appendix C: SI Prefixes Glossary Notation Acknowledgments for permission to reproduce copyrighted material Author Index Subject Index.
LC Subjects	Biomass energy. Biomass--Quality. Biomass conversion.
Other Subjects	Technology & Engineering / Chemical & Biochemical.
Notes	Includes bibliographical references and indexes.
Additional formats	Print version: Ptasinski, Krzysztof J., 1946- author. Efficiency of biomass energy Hoboken, New Jersey: Wiley, [2016] 9781118702109 (DLC) 2016020345

Efficiency of biomass energy: an exergy approach to biofuels, power, and biorefineries

LCCN	2016020345
Type of material	Book
Personal name	Ptasinski, Krzysztof J., 1946- author.

Main title	Efficiency of biomass energy: an exergy approach to biofuels, power, and biorefineries / Krzysztof J. Ptasinski.
Published/Produced	Hoboken, New Jersey: AIChE the global home of chemical engineers: Wiley, John Wiley & Sons, Inc., [2013]
Description	xxi, 756 pages; 29 cm
ISBN	9781118702109 (hardback)
LC classification	TP339 .P78 2016
Summary	"Details energy and exergy efficiencies of all major aspects of bioenergy systems Covers all major bioenergy processes starting from photosynthesis and cultivation of biomass feedstocks and ending with final bioenergy products, like power, biofuels, and chemicals Each chapter includes historical developments, chemistry, major technologies, applications as well as energy, environmental and economic aspects in order to serve as an introduction to biomass and bioenergy A separate chapter introduces a beginner in easy accessible way to exergy analysis and the similarities and differences between energy and exergy efficiencies are underlined Includes case studies and illustrative examples of 1st, 2nd, and 3rd generation biofuels production, power and heat generation (thermal plants, fuel cells, boilers), and biorefineries; Traditional fossil fuels-based technologies are also described in order to compare with the corresponding bioenergy systems."-- Provided by publisher.
LC Subjects	Biomass energy. Biomass--Quality. Biomass conversion.
Other Subjects	Technology & Engineering / Chemical & Biochemical.
Notes	Includes bibliographical references and indexes.
Additional formats	Online version: Ptasinski, Krzysztof J., 1946- author. Efficiency of biomass energy Hoboken,

New Jersey: Wiley, [2016] 9781119118152 (DLC) 2016024267

Endophytes: crop productivity and protection: volume 2

LCCN	2019747373
Type of material	Book
Main title	Endophytes: Crop Productivity and Protection: Volume 2 / edited by Dinesh K. Maheshwari, K. Annapurna.
Edition	1st ed. 2017.
Published/Produced	Cham: Springer International Publishing: Imprint: Springer, 2017.
Description	1 online resource (XI, 291 pages 70 illustrations) PDF
ISBN	9783319665443
Related names	Annapurna, K. editor. Maheshwari, Dinesh K. editor.
Summary	This book reviews the latest developments in our understanding of microbial endophytes and their potential applications in enhancing productivity and disease protection. It covers all the latest discoveries regarding endophytes, their interactions with plants and application in agricultural productivity and protection. Our understanding of endophytes has increased exponentially in recent decades. These microbes, such as fungi, bacteria, and actinobacteria, establish a symbiotic or parasitic association with plants. A better understanding of endophytic microorganisms may help to elucidate their functions and potential role in developing sustainable systems of crop production and improved protection against biotic stresses. Endophytes play a vital role in plant growth and health promotion. Endophytic bacteria are of agrobiological interest because they create host-endophyte relationships, which can open exciting prospects for newer biotechnological applications. Endophytes have also proven to be a beneficial and

sustainable alternative to agrochemicals due to their role in the biocontrol of pests and diseases. Further, endophytes are essential to the production of several secondary metabolites in grasses, in the process of gummosis in trees, and the production of useful metabolites such as alkaloids, pestaloside, cryptocandin, enfumafungin, subglutinols, et cetera for the host plant. They are also involved in the production of enzymes, biosurfactants, biocontrol agents and plant growth promoters. As such, it is imperative that we explore these products' industrial applications in the fields of biotechnology, pharmacy and agriculture. This volume will offers a valuable guidance for botanists, microbiologists, biotechnologists, molecular biologists, environmentalists, policymakers, conservationists, and those working for the protection of plant species of agricultural and medicinal importance.

Contents

1. Endophytic in Agrobiology: An Introduction - PART 1: Crop Productivity - 2. Role of Endophytic Bacteria on the Crop Productivity and Soil Health - 3. Plant Growth Promotion by Endophytic Bacteria in Non-Indigenous Crops - 4. Role of Endophytic Fungi in Agricultural Productivity - 5. Endophytic Bacteria: Role in Phosphate Solubilization - 6. Phytohormone Production from Edophytic Bacteria and Their Application in Plant Growth Promotion - 7. Beneficial Effects of Bacterial Endophytes on Forest Tree Species - 8. Role of Bacterial Endophytes in Plant Disease Control - 9. Endophytic Actinomycetes for Sustainable Agricultural Applications - PART 2: Metabolites Production - 10. Endophytic Fungi: A Remarkable Source of Biologically Active Scondary Metabolites - 11. Endophytes: Potential Source of Therpeuatically Important Secondary Metabolites of Plant Origin - 12. Endophytic Microorganisms: Source of Volatile Organic

	Compounds for Improvement of Plant Health - 13. Potential of Endophytic Lignin Degrading Fungi: Role on Lignocellulosic Biorefineries - 14. Conclusion - Index.
LC Subjects	Agriculture. Biotechnology. Microbial ecology. Plant biochemistry.
Other Subjects	Plant Biochemistry. Agriculture. Biotechnology. Microbial Ecology.
Additional formats	Print version: Endophytes: crop productivity and protection. 9783319665436 (DLC) 2017950012 Printed edition: 9783319665436 Printed edition: 9783319665450 Printed edition: 9783319882673
Series	Sustainable Development and Biodiversity, 2352-474X; 16 Sustainable Development and Biodiversity, 2352-474X; 16

Energy footprints of the bio-refinery, hotel, and building sectors

LCCN	2019751936
Type of material	Book
Main title	Energy Footprints of the Bio-refinery, Hotel, and Building Sectors / edited by Subramanian Senthilkannan Muthu.
Edition	1st ed. 2019.
Published/Produced	Singapore: Springer Singapore: Imprint: Springer, 2019.
Description	1 online resource (VII, 128 pages 22 illustrations, 15 illustrations in color.) PDF
ISBN	9789811324666
Related names	Muthu, Subramanian Senthilkannan. editor.
Summary	This book deals with the energy footprints of biorefineries and the hotel and buildings sector. It presents footprint case studies, which include

	background information, methodological frameworks, assessment checklists, calculation tools and techniques, applications, challenges and limitations. It also discusses the application of each indicator/framework in various industrial sectors and the associated challenges, along with outlooks for the future. Consumption and conservation of energy are key elements in any industry's sustainability strategy.
Contents	Energy Footprint of Biorefinery Schemes - A Technical Review on Methods and Tools for Evaluation of Energy Footprints, Impact on Buildings and Environment - Establishment of Electrical Energy Benchmarking Protocol for the Assessment of the Carbon Emission in Hotel Industry.
LC Subjects	Building. Buildings-Design and construction. Construction. Energy and state. Energy policy. Engineering, Architectural. Industrial management--Environmental aspects. Management. Natural resources. Sustainable development. Tourism.
Other Subjects	Energy Policy, Economics and Management. Building Construction and Design. Natural Resource and Energy Economics. Sustainability Management. Sustainable Development. Tourism Management.
Additional formats	Print version: Energy footprints of the bio-refinery, hotel, and building sectors. 9789811324659 (DLC) 2018953586 Printed edition: 9789811324659 Printed edition: 9789811324673 Printed edition: 9789811347795

Series	Environmental Footprints and Eco-design of Products and Processes, 2345-7651 Environmental Footprints and Eco-design of Products and Processes, 2345-7651

Extremophilic microbial processing of lignocellulosic feedstocks to biofuels, value-added products, and usable power

LCCN	2019746946
Type of material	Book
Main title	Extremophilic Microbial Processing of Lignocellulosic Feedstocks to Biofuels, Value-Added Products, and Usable Power / edited by Rajesh K. Sani, Navanietha Krishnaraj Rathinam.
Edition	1st ed. 2018.
Published/Produced	Cham: Springer International Publishing: Imprint: Springer, 2018.
Description	1 online resource (XII, 308 pages 46 illustrations, 32 illustrations in color.) PDF
ISBN	9783319744599
Related names	Krishnaraj Rathinam, Navanietha. editor. Sani, Rajesh K. editor.
Summary	This book presents a review and in-depth analyses of improved biotechnological processes emphasizing critical aspects and challenges of lignocellulosic biomass conversion into biofuels and value-added products especially using extremophiles and recombinant microorganisms. The book specifically comprises extremophilic production of liquid and gaseous biofuels (bioethanol, biobutanol, biodiesel, biohydrogen, and biogas) as well as value added products (e.g., single cell protein, hydrocarbons, lipids, exopolysaccharides, and polyhydroxyalkanoates). The book also provides the knowledge on how to develop safe, more efficient, sustainable, and economical integrated processes for enhanced conversion of lignocellulosic feedstocks to liquid and gaseous biofuels. Finally the book describes

	how to perform the techno-economical and life-cycle assessments of new integrated processes involving extremophiles. These modeling exercises are critical in addressing any deficiencies associated with the demonstration of an integrated biofuels and value-added products production process at pilot scale as well as demonstration on the commercialization scale.
Contents	Preface - Contributions - Acknowledgements - Bioprospecting of Extremophiles for Biotechnology Applications - Conversion of Lignocellulosic Feedstocks into Bioethanol using Extremophiles - Biobutanol production using Recombinant microorganisms/Extremophiles - Biodiesel from Extremophilic Microalgae - Hydrogen production from lignocellulosic feedstocks using extremophiles - Biogas from Extremophiles - Conversion of Lignocellulosic Feedstocks into Biogas: Upgrading and Utilization - Bioconversion of Methane for Value Added Products.-Exopolysaccharides Production from Extremophiles: The Chemical Structure and Their Bioactivities - Polyhydroxyalkanoates Production from Renewable and Waste Materials using Extremophiles/Recombinant Microbes - Rewiring Extremophilic Electrocatalytic Processes for Production of Biofuels and Value-added Compounds from Lignocellulosic Biomass - Integrated Consolidated Bioprocessing for Conversion of Lignocellulosic Feedstock to Biofuels and Value-added Bioproducts - Value added products from wastes using extremophiles in biorefineries - Process modelling, simulation and optimization tools - Index.
LC Subjects	Microbiology. Renewable energy sources.
Other Subjects	Microbiology. Applied Microbiology. Renewable and Green Energy.

Additional formats	Print version: Extremophilic microbial processing of lignocellulosic feedstocks to biofuels, value-added products, and usable power 9783319744575 (DLC) 2018943408 Printed edition: 9783030089955 Printed edition: 9783319744575 Printed edition: 9783319744582

Fermentation microbiology and biotechnology

LCCN	2018037045
Type of material	Book
Main title	Fermentation microbiology and biotechnology / edited by, Dr. E. M. T. El-Mansi (Lead editor), Dr. Jens Nielsen (Senior editor), Dr. David Mousdale (editor), Dr. Tony Allman, (editor) Dr. Ross Carlson (editor).
Edition	Fourth edition.
Published/Produced	Boca Raton: CRC Press, Taylor & Francis, 2019. ©2019
Description	xix, 419 pages: illustrations; 29 cm
ISBN	9781138581029 (hardback: alk. paper)
LC classification	TP248.27.M53 2019
Related names	El-Mansi, Mansi, editor. Nielsen, Jens, editor. Mousdale, David M., editor. Allman, Tony, editor. Carlson, Ross, editor.
Contents	Fermentation microbiology and biotechnology: central and modern concepts - Fermentation microbiology and biotechnology: an historical perspective / Mansi El-Mansi - Microbiology of industrial fermentation: central and modern concepts / Mansi El-Mansi - Fermentation kinetics: central and modern concepts / Jens Nielsen - Microbial synthesis of primary metabolites: current trends and future prospects / Arnold I. Demain and Sergio Sanchez - Microbial synthesis of secondary metabolites and strain improvement: current trends and future prospects / Sergio Sánchez, Beatriz

Ruiz-Villafán, Corina D. Ceapa and A.L. Demain - Flux control analysis and stoichiometric network modeling: current trends and future prospects / Mansi El-Mansi, Gregory Stephanopoulos, and Ross P. Carlson - Control of metabolite efflux in microbial cell factories: current advances and future prospects / Douglas B. Kell - Increasing tolerance and robustness of industrial microorganisms: current strategies and future prospects / Rui Pereira and Yun Chen - Engineering of microorganisms for production of pharmaceutical biomolecules / Adilson José da Silva, Bianca Barquera, and Mattheos A. Gl Koffas - Enzyme and cofactor engineering: current trends and future prospects in the pharmaceutical and fermentation industries / George N. Bennett and Ka-Yiu San - Conversion of renewable resources to biofuels and fine chemicals: current trends and future prospects / Muhammad Javed, Namdar Baghaei-Yazdi, Aristos Aristidou, and Brian S. Hartley - Biorefineries: industrial perspectives and challenges in primary and secondary metabolism / David Michael Mousdale - Solid-state fermentation of renewable resources: current trends and future prospects / Lalita Devi Gottumukkala, Kuniparambil Rajasree, Reeta Rani Singhania, Carlos Ricardo Soccol, and Ashok Pandey - Functional genomics: current trends, tools, and future prospects in the fermentation and pharmaceutical industries / Surendra K. Chikara, Toral Joshi, and Mahmoud Mostapha - Bioreactors: design, operation, and applications / Tony Allman - Biosensors in bioprocess monitoring and control: current trends and future prospects / Chris E. French - Immobilization and its applications in biotechnology: current trends and future prospects / Ronnie G. Willaert - Impact of microtitier plate based fermentations on bioprocess development: recent innovations and future prospects / Leonardo Rios-Solis, Patrick Wilk, and Halim Murni - Control and monitoring of industrial fermentations: an industrial perspective and recent trends / Gilles Roux, Cesar Arturo Aceves-Lara, Zetao Li, Boutaib Dahhou, and Craig J.L. Gershater.

Other Subjects	Industrial Microbiology Fermentation--physiology Microbiological Techniques
Notes	Includes bibliographical references and index.

Fermentation microbiology and biotechnology

LCCN	2019717951
Type of material	Book
Main title	Fermentation microbiology and biotechnology / edited by, Dr. E. M. T. El-Mansi (Lead editor), Dr. Jens Nielsen (Senior editor), Dr. David Mousdale (editor), Dr. Tony Allman, (editor) Dr. Ross Carlson (editor).
Edition	Fourth edition.
Published/Produced	Boca Raton: CRC Press, Taylor and Francis, 2019. ©2019
Description	1 online resource
ISBN	9780429015953 epub 9780429506987 general 9780429015960 pdf 9780429015946 mobi/kindle hardback: alk. paper
LC classification	TP248.27.M53
Related names	El-Mansi, Mansi, editor. Nielsen, Jens, editor. Mousdale, David M., editor. Allman, Tony, editor. Carlson, Ross, editor.
Contents	Fermentation microbiology and biotechnology: central and modern concepts - Fermentation microbiology and biotechnology: an historical perspective / Mansi El-Mansi - Microbiology of industrial fermentation: central and modern

concepts / Mansi El-Mansi - Fermentation kinetics: central and modern concepts / Jens Nielsen - Microbial synthesis of primary metabolites: current trends and future prospects / Arnold I. Demain and Sergio Sanchez - Microbial synthesis of secondary metabolites and strain improvement: current trends and future prospects / Sergio Sánchez, Beatriz Ruiz-Villafán, Corina D. Ceapa and A.L. Demain - Flux control analysis and stoichiometric network modeling: current trends and future prospects / Mansi El-Mansi, Gregory Stephanopoulos, and Ross P. Carlson - Control of metabolite efflux in microbial cell factories: current advances and future prospects / Douglas B. Kell - Increasing tolerance and robustness of industrial microorganisms: current strategies and future prospects / Rui Pereira and Yun Chen - Engineering of microorganisms for production of pharmaceutical biomolecules / Adilson José da Silva, Bianca Barquera, and Mattheos A. Gl Koffas - Enzyme and cofactor engineering: current trends and future prospects in the pharmaceutical and fermentation industries / George N. Bennett and Ka-Yiu San - Conversion of renewable resources to biofuels and fine chemicals: current trends and future prospects / Muhammad Javed, Namdar Baghaei-Yazdi, Aristos Aristidou, and Brian S. Hartley - Biorefineries: industrial perspectives and challenges in primary and secondary metabolism / David Michael Mousdale - Solid-state fermentation of renewable resources: current trends and future prospects / Lalita Devi Gottumukkala, Kuniparambil Rajasree, Reeta Rani Singhania, Carlos Ricardo Soccol, and Ashok Pandey - Functional genomics: current trends, tools, and future prospects in the fermentation and pharmaceutical industries / Surendra K. Chikara, Toral Joshi, and Mahmoud Mostapha - Bioreactors: design, operation, and applications /

	Tony Allman - Biosensors in bioprocess monitoring and control: current trends and future prospects / Chris E. French - Immobilization and its applications in biotechnology: current trends and future prospects / Ronnie G. Willaert - Impact of microtitier plate based fermentations on bioprocess development: recent innovations and future prospects / Leonardo Rios-Solis, Patrick Wilk, and Halim Murni - Control and monitoring of industrial fermentations: an industrial perspective and recent trends / Gilles Roux, Cesar Arturo Aceves-Lara, Zetao Li, Boutaib Dahhou, and Craig J.L. Gershater.
Other Subjects	Industrial Microbiology Fermentation--physiology Microbiological Techniques
Notes	Includes bibliographical references and index.
Additional formats	Print version: Fermentation microbiology and biotechnology Boca Raton: CRC Press, Taylor & Francis, 2019. 9781138581029 (hardback: alk. paper) (DLC) 2018037045

Food waste reduction and valorisation: sustainability assessment and policy analysis

LCCN	2019749347
Type of material	Book
Main title	Food Waste Reduction and Valorisation: Sustainability Assessment and Policy Analysis / edited by Piergiuseppe Morone, Franka Papendiek, Valentina Elena Tartiu.
Edition	1st ed. 2017.
Published/Produced	Cham: Springer International Publishing: Imprint: Springer, 2017.
Description	1 online resource (VIII, 327 pages 50 illustrations, 35 illustrations in color.) PDF
ISBN	9783319500881
Related names	Morone, Piergiuseppe. editor. Papendiek, Franka. editor.

Tartiu, Valentina Elena. editor.

Summary

This book adds a new dimension to the sustainability assessment of food waste reduction and valorisation: policy analysis. Featuring a transdisciplinary analysis by key experts in the field, it identifies the drivers of change in food-waste reduction and valorisation technologies by looking, for example, at the regulatory framework and at policy actions undertaken by local and global actors. The book explores the development of regulations and policies for food-waste prevention, management, and valorisation at a global as well as European Union level. It also discusses the notion of food waste in legal terms and investigates the effects of the lack of a standard, universal definition of food waste on the efficient use of by-products, promising processes and products for technological and commercial exploitation. Utilising mathematical mapping methods to assess food consumption impacts and providing supply chain models that allow the testing of consumption scenarios, the book goes on to discuss a series of emerging technologies (tested at lab scale and/ or pilot scale) and opportunities for the valorisation of food waste.

Contents

technological scale up and commercial exploitation - Techno-economic study and environmental assessment of food waste based biorefinery - Techno-economic evaluation of refining of food supply chain wastes for the production of chemicals and biopolymers - Bio-based Economy: Policy Framework and Foresight Thinking - Bringing a Sharing Economy Approach into the Food Sector: The Potential of Food Sharing for Reducing Food Waste - Defining the Meaning of Food Waste as a Matter of Urgency - Waste Reduction in Fresh Food Supply Chains: an Operations Research Approach - Participatory Planning in Organic Solid Waste Management: A Backcasting Approach - The role of social networks in the diffusion of bio-waste products: the case of mulching films derived from organic waste in Province of Foggia - Grassroots innovations and the transition towards sustainability: Tackling the food waste challenge.

LC Subjects Climate change.
Energy and state.
Energy policy.
Sustainable development.
Waste management.

Other Subjects Energy Policy, Economics and Management.
Climate Change/Climate Change Impacts.
Sustainable Development.
Waste Management/Waste Technology.

Additional formats Print version: Food waste reduction and valorisation. 9783319500874 (DLC) 2016959520
Printed edition: 9783319500874
Printed edition: 9783319500898
Printed edition: 9783319843148

Food waste to valuable resources: applications and management

LCCN 2020951302

Type of material Book

Main title Food waste to valuable resources: applications and management / Rajesh Banu.
Edition 1.
Published/Produced San Diego: Elsevier Inc, 2020.
ISBN 9780128183533 (paperback)
(ebook)
Related names Banu, Rajesh, editor.
Summary "Food Waste to Valuable Resources: Applications and Management compiles current information pertaining to food waste, placing particular emphasis on the themes of food waste management, biorefineries, valuable specialty products and technoeconomic analysis. Following its introduction, this book explores new valuable resource technologies, the bioeconomy, the technoeconomical evaluation of food-waste-based biorefineries, and the policies and regulations related to a food-waste-based economy. It is an ideal reference for researchers and industry professionals working in the areas of food waste valorization, food science and technology, food producers, policymakers and NGOs, environmental technologists, environmental engineers, and students studying environmental engineering, food science, and more"-- Provided by publisher.

Fungal biorefineries
LCCN 2018946100
Type of material Book
Main title Fungal biorefineries / Sachin Kumar.
Published/Produced New York, NY: Springer Science+Business Media, 2018.
ISBN 9783319903781

Fungal biorefineries
LCCN 2019769479
Type of material Book

Main title	Fungal Biorefineries / edited by Sachin Kumar, Pratibha Dheeran, Mohammad Taherzadeh, Samir Khanal.
Edition	1st ed. 2018.
Published/Produced	Cham: Springer International Publishing: Imprint: Springer, 2018.
Description	1 online resource (XIV, 246 pages 32 illustrations, 27 illustrations in color.) PDF
ISBN	9783319903798
Related names	Dheeran, Pratibha, 1980- editor. Khanal, Samir, editor. Kumar, Sachin, 1981- editor. Taherzadeh, Mohammad, editor.
Summary	This book covers the applications of fungi used in biorefinery technology. As a great many different varieties of fungal species are available, the text focuses on the various applications of fungi for production of useful products including organic acids (lactic, citric, fumaric); hydrolytic enzymes (amylase, cellulases, xylanases, ligninases, lipases, pectinases, proteases); advanced biofuels (ethanol, single cell oils); polyols (xylitol); single cell protein (animal feed); secondary metabolites; and much more.
Contents	Preface - Chapter 1 Role of Fungi in Biorefinery: A Perspective Kanika Chowdhary, Usha Prasad, Satyawati Sharma - Chapter 2 Production of Biofuels from Biomass by Fungi Gail Joseph, Lijun Wang - Chapter 3 White-rot Fungal Xylanases for Applications in Pulp and Paper industry Shalini Singh - Chapter 4 Fungal Enzymes Applied to Industrial Processes for Bioethanol Production Cecilia Laluce, Kelly J. Dussán, Marisa V. Capella - Chapter 5 Economical Lactic Acid Production and Optimization Strategies Sheelendra M. Bhatt - Chapter 6 Exploiting Innate and Imported Fungal Capacity for Xylitol Production Shaik Jakeer - Chapter 7 Insights into Fungal Xylose Reductases

	and its Application in Xylitol Production Yogita Lugani, Balwinder Singh Sooch - Chapter 8 Single Cell Oils (SCOs) of Oleaginous Filamentous Fungi as a Renewable Feedstock: A Biodiesel Biorefinery Approach Mahesh Khot, Gouri Katre, Smita Zinjarde, Ameeta RaviKumar - Chapter 9 Fungal Biorefinery for the Production of Single Cell Oils as Advanced Biofuels Abu Yousuf, Baranitharan Ethiraj, Maksudur Rahman Khan, Domenico Pirozzi - Chapter 10 Production of Single Cell Protein (SCP) from Vinasse Ernesto Acosta Martínez, Jéssica Ferreira dos Santos, Geiza Suzart Araujo, Sílvia Maria Almeida de Souza, Rita de Cássia Lacerda Brambilla Rodrigues, and Eliana Vieira Canettieri - Index.
LC Subjects	Microbiology. Mycology. Plant anatomy. Plant development. Plant genetics. Plant systematics. Plant taxonomy.
Other Subjects	Mycology. Microbiology. Plant Anatomy/Development. Plant Genetics and Genomics. Plant Systematics/Taxonomy/Biogeography.
Additional formats	Print version: Fungal biorefineries 9783319903781 (DLC) 2018946100 Printed edition: 9783030080020 Printed edition: 9783319903781 Printed edition: 9783319903804
Series	Fungal Biology, 2198-7777 Fungal Biology, 2198-7777

Furfural: an entry point of lignocellulose in biorefineries to produce renewable chemicals, polymers, and biofuels

LCCN	2017054509
Type of material	Book

Main title	Furfural: An Entry Point of Lignocellulose in Biorefineries to Produce Renewable Chemicals, polymers, and biofuels / editors, Manuel Löpez Granados (Institute of Catalysis and Petrochemistry (CSIC), Spain), David Martin Alonso (University of Wisconsin-Madison, USA).
Published/Produced	New Jersey: World Scientific, [2018]
Description	xxiv, 357 pages; 24 cm.
ISBN	9781786344861 (hc; alk. paper)
LC classification	TP680 .F87 2018
Related names	López Granados, Manuel, editor.
	Alonso, David Martin, editor.
LC Subjects	Furfural.
	Vegetable oils.
	Biomass energy.
	Lignocellulose.
	Agricultural wastes.
Notes	Includes bibliographical references and index.
Series	Sustainable chemistry series; volume 2

Hemicelluloses and lignin in biorefineries

LCCN	2017035505
Type of material	Book
Personal name	Wertz, Jean-Luc, author.
Main title	Hemicelluloses and lignin in biorefineries / by Jean-Luc Wertz, Magali Deleu, Séverine Coppée, Aurore Richel.
Published/Produced	Boca Raton: CRC Press, Taylor & Francis Group, [2018]
Description	xxii, 308 pages: illustrations (some color); 24 cm.
ISBN	9781138720985 (hardbook: acid-free paper)
LC classification	TP339 .W47 2018
LC Subjects	Cellulosic ethanol.
	Lignocellulose--Biotechnology.
	Ethanol as fuel.
Notes	Includes index.
Series	Green chemistry and chemical engineering

Hydrothermal processing in biorefineries: production of bioethanol and high added-value compounds of second and third generation biomass

LCCN	2019761369
Type of material	Book
Main title	Hydrothermal Processing in Biorefineries: Production of Bioethanol and High Added-Value Compounds of Second and Third Generation Biomass / edited by Héctor A. Ruiz, Mette Hedegaard Thomsen, Heather L. Trajano.
Edition	1st ed. 2017.
Published/Produced	Cham: Springer International Publishing: Imprint: Springer, 2017.
Description	1 online resource (XXV, 511 pages 159 illustrations, 79 illustrations in color.) PDF
ISBN	9783319564579
Related names	Hedegaard Thomsen, Mette. editor. Ruiz, Héctor A. editor. Trajano, Heather L. editor.
Summary	The biorefinery, integration of processes and technologies for biomass conversion, demands efficient utilization of all components. Hydrothermal processing is a potential clean technology to convert raw materials such as lignocellulosic and aquatic biomass into bioenergy and high added-value compounds. This book aims to show fundamental concepts and key technological developments that enabled industrial application of hydrothermal processing. The scope of this book is primarily for scientists working in the biorefinery field as well as engineers from industry and potential investors in biofuels. Therefore, the information in this book will provide an overview of this technology applied to lignocellulosic materials and aquatic biomass, and especially new knowledge. Critically, this book brings together experts in the application of hydrothermal processes on lignocellulosic and

aquatic biomass. Focuses on hydrothermal process as a fundamental step in the biorefinery; Covers the use of both marine and lignocellulosic biomass; Features information on tools for industrial scale-up and methods for techno-economic analysis.

Contents

Introduction Book - Effect of Hydrothermal Pretreatment on Lignin and Antioxidant Activity - Effect of Hydrothermal Processing on Hemicellulose Structure - Response of Biomass Species to Hydrothermal Pretreatment - Kinetic Modeling, Operational Conditions and Biorefinery Products from Hemicellulose: Depolymerization and Solubilization during Hydrothermal Processing - Combined Severity Factor for Predicting Sugar Recovery in Acid Catalyzed Pretreatment Followed by Enzymatic Hydrolysis - Hydrothermal Pretreatment of Lignocellulosic Biomass for Bioethanol Production - Hydrothermal Pretreatment: Process Modeling and Economic Assessment within the Framework of Biorefinery Processes - Bioethanol Production from Pretreated Solids Using Hydrothermal Processing - Production and Emerging Applications of Bioactive Oligosaccharides from Biomass Hemicelluloses by Hydrothermal Processing - Production of Hemicellulases, Xylitol and Furan from Hemicellulose Hydrolysates Using Hydrothermal Pretreatment - Steam Explosion as a Hydrothermal Pretreatment in the Biorefinery Concept - Adaptation of Severity Factor Model According to the Operating Parameters Variations which occur during Steam Explosion Process - Hydrothermal Pre-treatment Using Supercritical CO2 in the Biorefinery Context - Scale-up Hydrothermal Pre-treatment of Sugarcane Bagasse and Straw for Second-Generation Ethanol Production - Pilot Plant Design and Operation Using a Hydrothermal Pretreatment: Bioenercel Experience - Techno-economic Aspects in the

	Evaluation of Biorefineries for Production of Second Generation Bioethanol - Minimizing Precipitated Lignin Formation and Maximizing Monosugar Concentration by Formic Acid Reinforced Hydrolysis of Hardwood Chips - Microwave-assisted Hydrothermal Processing of Seaweed Biomass - Hydrothermal Processes for Extraction of Macroalgae high Value added Compounds - Hydrothermal Processing of Microalgae.
LC Subjects	Biochemical engineering. Heat engineering. Heat transfer. Mass transfer. Renewable energy sources. Thermodynamics.
Other Subjects	Renewable and Green Energy. Biochemical Engineering. Engineering Thermodynamics, Heat and Mass Transfer. Renewable and Green Energy.
Additional formats	Print version: Hydrothermal processing in biorefineries. 9783319564562 (DLC) 2017941184 Printed edition: 9783319564562 Printed edition: 9783319564586 Printed edition: 9783319859163

Hydrothermal processing in biorefineries

LCCN	2017941184
Type of material	Book
Main title	Hydrothermal processing in biorefineries.
Published/Produced	New York, NY: Springer International Pub., 2017.
ISBN	9783319564562

Life-cycle assessment of biorefineries

LCCN	2016461764
Type of material	Book
Main title	Life-cycle assessment of biorefineries / [edited by] Edgard Gnansounou, Ashok Pandey.

Published/Produced	Amsterdam, Netherlands: Elsevier, [2017] ©2017
Description	x, 312 pages: illustrations; 25 cm
ISBN	0444635858 9780444635853
LC classification	TP248.25.B55 L54 2017
Related names	Gnansounou, Edgard, editor. Pandey, Ashok, editor.
LC Subjects	Bioreactors. Biomass--Refining. Product life cycle.
Notes	Includes bibliographical references and index.

Macroalgal biorefineries for the blue economy

LCCN	2020034551
Type of material	Book
Personal name	Golberg, Alexander, author.
Main title	Macroalgal biorefineries for the blue economy / Alexander Golberg, Tel Aviv University, Israel; Arthur Nils Robin,Tel Aviv University, Israel; Meiron Zollmann,Tel Aviv University, Israel; Hadar Traugott, Tel Aviv University, Israel; Ruslana Rachel Palatnik, University of Haifa, Israel; Alvaro Israel, Israel Oceanographic & Limnological Research Institute, Israel.
Published/Produced	New Jersey: World Scientific, [2020]
ISBN	9789811224287 (hardcover) (ebook for institutions) (ebook for individuals)
LC classification	TP339 .G64 2020
Related names	Robin, Arthur Nils, author. Zollmann, Meiron, author. Traugott, Hadar, author. Palatnik, Ruslana Rachel, author. Israel, Alvaro, author.
Summary	"This unique compendium provides an insight into the role of emerging marine biorefineries based on macroalgae (seaweeds) in the development of a sustainable use of ocean resources for economic

growth-blue economy. The useful reference text shows a complete picture on the motivation to develop seaweed technologies, and how the combination of biology, cultivation technologies and downstream processing with economics can address the social challenges through the blue growth"-- Provided by publisher.

Contents	Why going to seagriculture? The need for new sources of biomass for bioeconomy and what can seas offer to humanity - Blue economy and marine biorefinery - Marine macroalgae - What is macroalgae biorefinery? - Environmental impacts of seaweed aquaculture - The marine biomass feedstock - Downstream processing - Examples of applications - Economics of macroalgae biorefineries.
LC Subjects	Algal biofuels. Marine algae--Refining. Marine algae--Biotechnology. Marine algae industry. Marine algae culture. Sustainable aquaculture. Biochemical engineering.
Notes	Includes bibliographical references and index.

Meeting policy challenges for a sustainable bioeconomy.

LCCN	2018381315
Type of material	Book
Main title	Meeting policy challenges for a sustainable bioeconomy.
Published/Produced	Paris: OECD, [2018] ©2018
Description	195 pages: illustrations (some color), color maps; 28 cm
Links	Available to Stanford-affiliated users. https://doi.org/ 10.1787/9789264292345-en OECD iLibrary
ISBN	9264292330 9789264292338

LC classification	QH705 .M43 2018
Related names	Organisation for Economic Co-operation and Development.
Contents	The bioeconomy concept: then and now - Reconciling food and industrial needs for biomass - Measuring biomass potential and sustainability - Biotechnology and biomass sustainability - What is a biorefinery: definitions, classification and general models - Financing biorefineries - Biowaste biorefining - Developments in bio-based production - Enabling bio-based materials policy - Metabolic engineering and synthetic biology for a bioeconomy - Education and training for industrial biotechnology.
LC Subjects	Biology, Economic. Sustainable development. Sustainability. Biotechnology--Economic aspects. Biotechnology--Government policy. Biomass energy. Biomass--Refining.
Notes	Includes bibliographical references.

Microalgae cultivation for biofuels production

LCCN	2020277593
Type of material	Book
Main title	Microalgae cultivation for biofuels production / edited by Abu Yousuf.
Published/Produced	San Diego, CA: Academic Press, an imprint of Elsevier, [2020] ©2020
Description	xx, 361: color illustrations; 24 cm
ISBN	0128175362 9780128175361
Related names	Yousuf, Abu, editor.
Summary	"Microalgae Cultivation for Biofuels Production explores the technological opportunities and challenges involved in producing economically competitive algal-derived biofuel. The book

discusses efficient methods for cultivation, improvement of harvesting and lipid extraction techniques, optimization of conversion/production processes of fuels and co-products, the integration of microalgae biorefineries to several industries, environmental resilience by microalgae, and a techno-economic and lifecycle analysis of the production chain to gain maximum benefits from microalgae biorefineries." - Provided by publisher.

Contents Fundamentals of microalgae cultivation / Abu Yousuf - Microalgae cultivation systems / Gaetano Zuccaro, Abu Yousuf, Antonino Pollio and Jean-Philippe Steyer - Microalgae cultivation and photobioreactor design / Wenguang Zhou, Qian Lu, Pei Han and Jingjing Li - Mixotrophic cultivation: biomass and biochemical biosynthesis for biofuel production / Tan Kean Meng, Mohd Asyraf Kassim and Benjamas Cheirsilp - Oleaginous microalgae cultivation for biogas upgrading and phytoremediation of wastewater / Sirasit Srinuanpan, Benjamas Cheirsilp and Mohd Asyraf Kassim - Efficient harvesting of microalgal biomass and direct conversion of microalgal lipids into biodiesel / Benjamas Cheirsilp, Sirasit Srinuanpan and Yohanis Irenius Mandik - Membrane technology for microalgae harvesting / Jia Jia Leam, Muhammad Roil Bilad, Yusuf Wibisono, Mohd Dzul Hakin Wirzal and Iftikhar Ahmed - Processing of microalgae to biofuels / Emmanuel Manirafasha, Kailin Jiao, Xianhai Zeng, Yuanchao Xu, Xing Tang, Yong Sun, Lu Lin, Theophile Murwanashyaka, Theoneste Ndikubwimana, Keju Jing and Yinghua Lu - Microalgal cell disruption and lipid extraction techniques for potential biofuel production / Ahasanul Karim, M. Amirul Islam, Zaied Bin Khalid, Che Ku Mohammad Faizal, Md. Maksudur Rahman Khan and Abu Yousuf - Conversion of microalgae biomass to biofuels / Min-Yee Choo,

Lee Eng Oi, Tau Chuang Ling, Eng-Poh Ng, Hwei Voon Lee and Joon Ching Juan - Microalgal biorefinery / Eleni Koutra, Panagiota Tsafrakidou, Myrsini Sakarika and Michael Kornaros - Microalgal biorefineries for industrial products / Sabeela Beevi Ummalyma, Dinabandhu Sahoo and Ashok Pandey - Biorefinery of microalgae for nonfuel products / Ambreen Aslam, Tahir Fazal, Qamar Uz Zaman, Ali Shan, Fahad Rehman, Javed Iqbal, Naim Rashid and Muhammad Saif Ur Rehman - Recent trends in strain improvement for production of biofuels from microalgae / S.V. Vamsi Bharadwaj, Shristi Ram, Imran Pancha and Sandhya Mishra - Microalgae to biogas: microbiological communities involved / Olivia Córdova and Rolando Chamy - Microalgae-based biofuel production using low-cost nanobiocatalysts / S. Vijayalakshmi, M. Anand and J. Ranjitha - Biosynthesis of nanomaterials using algae / Ashiqur Rahman, Shishir Kumar and Tabish Nawaz - Life cycle assessment and techno-economic analysis of algal biofuel production / Dongyan Mu, Chunhua Xin and Wenguang Zhou - Environmental resilience by microalgae / M.L. Serejo, M. Franco Morgado, D. García, A. González-Sánchez, H.O. Méndez-Acosta and A. Toledo-Cervantes - Microalgae-based remediation of wastewaters / Myrsini Sakarika, Eleni Koutra, Panagiota Tsafrakidou, Antonia Terpou and Michael Kornaros - Resource recovery from waste streams using microalgae: opportunities and threats / Naim Rashid, Thinesh Selvaratnam and Won-Kun Park.

LC Subjects Microalgae--Biotechnology.
Biomass energy.

Notes Includes bibliographical references and index.

Microbial enzymes in bioconversions of biomass

LCCN 2019742581

Type of material	Book
Main title	Microbial Enzymes in Bioconversions of Biomass / edited by Vijai Kumar Gupta.
Edition	1st ed. 2016.
Published/Produced	Cham: Springer International Publishing: Imprint: Springer, 2016.
Description	1 online resource (VIII, 347 pages 50 illustrations, 20 illustrations in color.) PDF
ISBN	9783319436791
Related names	Gupta, Vijai Kumar. editor.
Summary	This volume summarizes our current knowledge on different biomass-converting enzymes and their potential use in converting biomass into simple sugar to generate bioenergy and other value added co-/by products. It consists of 13 chapters and is divided into three parts: cellulases; hemicellulases; and lignocellulose oxidoreductases. Written by international experts, the contributions offer clear and concise descriptions of both standard and new technologies. It is an invaluable reference resource for undergraduates, post-graduates, researchers and practitioners in the field of microbial enzymes for biofuel and biorefinery applications.
Contents	Microbial enzymes for conversion of biomass to bioenergy - Cellobiohydrolases - Role, Mechanism and Recent developments - Endo - 1, 4 - [beta] - glucanases: role, applications and recent developments - The Role and Applications of [beta]-Glucosidases in Biomass Degradation and Bioconversion - Role and applications of feruloyl esterases in biomass bioconversion - Endo-[beta]-1,4-xylanase: An overview of recent developments - Microbial xylanases: Sources, types and their applications - Mannanase - The Role and Applications of Xyloglucan Hydrolase in Biomass Degradation/Bioconversion - Role of mushroom Mn-oxidizing peroxidases in biomass conversion - Role and application of versatile peroxidase (VP)

	for utilizing lignocellulose in biorefineries - Fungal Aryl-alcohol Oxidase in Lignocellulose Degradation and Bioconversion - Monosaccharide oxygenase.
LC Subjects	Energy systems. Enzymology. Renewable energy sources.
Other Subjects	Renewable and Green Energy. Energy Systems. Enzymology.
Additional formats	Print version: Microbial enzymes in bioconversions of biomass. 9783319436777 (DLC) 2016947366 Printed edition: 9783319436777 Printed edition: 9783319436784 Printed edition: 9783319828848
Series	Biofuel and Biorefinery Technologies, 2363-7609; 3 Biofuel and Biorefinery Technologies, 2363-7609; 3

Petrochemistry: petrochemical processing, hydrocarbon technology, and green engineering

LCCN	2019039112
Type of material	Book
Personal name	Bajus, Martin, 1943- author.
Main title	Petrochemistry: petrochemical processing, hydrocarbon technology, and green engineering / Martin Bajus (Slovak University of Technology, Faculty of Chemical & Food Technology, Bratislava).
Edition	First edition.
Published/Produced	Hoboken, NJ: John Wiley & Sons, [2019]
Description	1 online resource
ISBN	9781119647836 (epub) 9781119647805 (adobe pdf) (hardback)
LC classification	TP692.3

Contents	The chemical technology - Current trend in green hydrocarbon technology - Energy technology - Source of hydrocarbons - Links with natural gas, crude oil, and petroleum refineries - Hydrocarbon technology, trends, and outlook in petrochemistry - Pillar of petrochemistry: production of lower alkenes - Pillar of petrochemistry: other sources of lower alkenes - Pillar of petrochemistry: petrochemicals from C_2-C_3 alkenes - Pillar of petrochemistry: production of BTX aromatics - Pillar of petrochemistry: chemicals from BTX aromatics - Pillar of petrochemistry: C1 technologies - Hydrogen technologies - Biorefineries - Recycling technologies - Microchannel technologies and nanotechnology.
LC Subjects	Petroleum chemicals. Petroleum--Refining. Alkenes. Chemistry, Technical.
Notes	Includes bibliographical references and index.
Additional formats	Print version: Bajus, Martin, 1943- Petrochemistry First edition. Hoboken, NJ: John Wiley & Sons, [2019] 9781119647768 (DLC) 2019039111

Petrochemistry: petrochemical processing, hydrocarbon technology, and green engineering

LCCN	2019039111
Type of material	Book
Personal name	Bajus, Martin, 1943- author.
Main title	Petrochemistry: petrochemical processing, hydrocarbon technology, and green engineering / Martin Bajus (Slovak University of Technology, Faculty of Chemical & Food Technology, Bratislava).
Edition	First edition.
Published/Produced	Hoboken, NJ: John Wiley & Sons, [2019]
ISBN	9781119647768 (hardback) (adobe pdf) (epub)

LC classification TP692.3 .B35275 2019

Contents The chemical technology - Current trend in green hydrocarbon technology - Energy technology - Source of hydrocarbons - Links with natural gas, crude oil, and petroleum refineries - Hydrocarbon technology, trends, and outlook in petrochemistry - Pillar of petrochemistry: production of lower alkenes - Pillar of petrochemistry: other sources of lower alkenes - Pillar of petrochemistry: petrochemicals from C_2-C_3 alkenes - Pillar of petrochemistry: production of BTX aromatics - Pillar of petrochemistry: chemicals from BTX aromatics - Pillar of petrochemistry: C1 technologies - Hydrogen technologies - Biorefineries - Recycling technologies - Microchannel technologies and nanotechnology.

LC Subjects Petroleum chemicals.
Petroleum--Refining.
Alkenes.
Chemistry, Technical.

Notes Includes bibliographical references and index.

Additional formats Online version: Bajus, Martin, 1943- Petrochemistry First edition. Hoboken: Wiley, 2020. 9781119647805 (DLC) 2019039112

Platform chemical biorefinery: future green chemistry

LCCN 2016946411

Type of material Book

Main title Platform chemical biorefinery: future green chemistry / edited by Satinder Kaur Brar, Institut national de la recherche scientifique Centre - Eau Terre Environnement (INRS-ETE), Quebec, Canada; Saurabh Jyoti Sarma, Institut national de la recherche scientifique Centre - Eau Terre Environnement (INRS-ETE), Quebec, Canada; Kannan Pakshirajan, Department of Biosciences and Bioengineering, Indian Institute of Technology Guwahati, Guwahati, Asam, India.

Published/Created Amsterdam; Boston: Elsevier, [2016]

Published/Produced	©2016
Description	xiv, 513 pages: illustrations; 24 cm
ISBN	9780128029800; (hbk.)
LC classification	TP248.B55 P53 2016
Cover title	Platform chemical biorefinery, future green industry
Related names	Brar, Satinder Kaur, editor.
	Sarma, Saurabh Jyoti, editor.
	Pakshirajan, Kannan, editor.
Summary	"Platform Chemical Biorefinery: Future Green Chemistry provides information on three different aspects of platform chemical biorefinery. The book first presents a basic introduction to the industry beneficial for university students, then provides engineering details of existing or potential platform chemical biorefinery processes helpful to technical staff of biorefineries. Finally, the book presents a critical review of the entire platform chemical biorefinery process, including extensive global biorefinery practices and their potential environmental and market-related consequences. Platform chemicals are building blocks of different valuable chemicals. The book evaluates the possibility of renewable feedstock-based platform chemical production and the fundamental challenges associated with this objective. Thus, the book is a useful reference for both academic readers and industry technical workers. The book guides the research community working in the field of platform chemical biorefinery to develop new pathways and technologies in combination with their market value and desirability"--Page 4 of cover.
LC Subjects	Biomass chemicals--Technological innovations.
	Biomass--Refining.
	Biomass conversion.
	Biomass energy--Technological innovations.
	Green chemistry.
Notes	Includes bibliographical references and index.

Production of biofuels and chemicals from lignin

LCCN	2019742682
Type of material	Book
Main title	Production of Biofuels and Chemicals from Lignin / edited by Zhen Fang, Richard L. Smith, Jr.
Edition	1st ed. 2016.
Published/Produced	Singapore: Springer Singapore: Imprint: Springer, 2016.
Description	1 online resource (XV, 435 pages 152 illustrations, 56 illustrations in color.) PDF
ISBN	9789811019654
Related names	Fang, Zhen. editor. Smith, Jr., Richard L. editor.
Summary	This book provides state-of-the-art reviews, current research on and the prospects of lignin production, biological, thermal and chemical conversion methods, and lignin technoeconomics. Fundamental topics related to lignin chemistry, properties, analysis, characterization, and depolymerization mechanisms, as well as enzymatic, fungal and bacterial degradation methods are covered. The book also examines practical topics related to technologies for lignin and ultra-pure lignin recovery, activated carbon, carbon fiber production and materials, and addresses the biological conversion of lignin with fungi, bacteria or enzymes to produce chemicals, along with chemical, catalytic, thermochemical and solvolysis conversion methods. Lastly, it presents a case study on practical polyurethane foam production using lignin. Lignin has a bright future and will be an essential feedstock for producing renewable chemicals, biofuels and value-added products. Offering comprehensive information on this promising material, the book represents a valuable resource for students, researchers, academicians and industrialists in the field of biochemistry and energy.

Contents	Part I: Lignin and Its Production (Chapters 1-3) - Part II: Biological Conversion (Chapters 4-6) - Part III: Chemical Conversion (Chapters 7-12) - Part IV: Techno-economics.
LC Subjects	Biochemistry. Forestry. Renewable energy sources.
Other Subjects	Biochemistry, general. Forestry. Renewable and Green Energy.
Additional formats	Print version: Production of biofuels and chemicals from lignin 9789811019647 (DLC) 2016951486 Printed edition: 9789811019647 Printed edition: 9789811019661 Printed edition: 9789811094897
Series	Biofuels and Biorefineries, 2214-1537 Biofuels and Biorefineries, 2214-1537

Production of biofuels and chemicals with bifunctional catalysts

LCCN	2019747308
Type of material	Book
Main title	Production of Biofuels and Chemicals with Bifunctional Catalysts / edited by Zhen Fang, Richard L. Smith Jr., Hu Li.
Edition	1st ed. 2017.
Published/Produced	Singapore: Springer Singapore: Imprint: Springer, 2017.
Description	1 online resource (XVI, 391 pages 244 illustrations, 36 illustrations in color.) PDF
ISBN	9789811051371
Related names	Fang, Zhen. editor. Li, Hu. editor. Smith Jr., Richard L. editor.
Summary	Bifunctional catalysts integrate sequential catalytic steps to efficiently transform raw materials into chemical products. Bifunctional catalysts avoid by-product formation and costly separation steps so that their design and application are seen as the next

generation of chemical technology. This text provides state-of-the-art reviews, current research, prospects and challenges of the production of biofuels and chemicals such as furanic biofuels, biodiesel, carboxylic acids, polyols and others from lignocellulosic biomass, furfurals, syngas and [gamma]-valerolactone with bifunctional catalysts, including catalytic, electrocatalytic, and combined biological and chemical catalysis processes. The bifunctionality of catalytic materials is a concept of not only using multifunctional solid materials as activators, but in such a way that the catalytic materials have synergistic characteristics that promote a cascade of transformations which reduces or eliminates reaction and separation steps. The book will be useful for students, researchers, academicians and industrialists in the field of biotechnology, biochemistry and energy.

LC Subjects Biomedical engineering.
Biotechnology.
Renewable energy sources.

Other Subjects Biomedical Engineering/Biotechnology.
Biotechnology.
Renewable and Green Energy.

Additional formats Print version: Production of biofuels and chemicals with bifunctional catalysts 9789811051364 (DLC) 2017961301
Printed edition: 9789811051364
Printed edition: 9789811051388
Printed edition: 9789811353291

Series Biofuels and Biorefineries, 2214-1537; 8
Biofuels and Biorefineries, 2214-1537; 8

Production of materials from sustainable biomass resources

LCCN 2019763906

Type of material Book

Main title Production of Materials from Sustainable Biomass Resources / edited by Zhen Fang, Richard L. Smith, Jr, Xiao-Fei Tian.

Edition	1st ed. 2019.
Published/Produced	Singapore: Springer Singapore: Imprint: Springer, 2019.
Description	1 online resource (XVIII, 401 pages 121 illustrations, 47 illustrations in color.) PDF
ISBN	9789811337680
Related names	Fang, Zhen, editor. Smith, Jr, Richard L, editor. Tian, Xiao-Fei, editor.
Summary	This book presents a collection of studies on state-of-art techniques developed specifically for lignocellulose component derivation, and for the production of functional materials, composite polymers, carbonaceous biocatalysts, and pellets from lignocellulosic biomass, with an emphasis on using sustainable chemistry and engineering to develop innovative materials and fuels for practical application. Technological strategies for the physical processing or biological conversion of biomass for material production are also presented. All chapters were contributed by respected experts in the field from around the globe, providing a broad range of perspectives on cutting-edge applications. The book offers an ideal reference guide for academic researchers and industrial engineers in the fields of natural renewable materials, biorefinery of lignocellulose, biofuels and environmental engineering. It can also be used as a comprehensive reference source for university students in chemical engineering, material science and environmental engineering.
Contents	Part 1: Isolation and Purification of Lignocellulose Components - Chapter 1. Isolation, Purification, and Potential Applications of Xylan - Part 2: Composite Polymers Derived from Lignin and Cellulose - Chapter 2. Development of Lignin-Based Antioxidants for Polymers - Chapter 3. Nanocellulose Applications in Papermaking - Part

	3: Functional Materials Derived from Cellulose and Lignocelluloses - Chapter 4. Recent Advances in Cellulose Chemistry and Potential Applications - Chapter 5. Production, Characterization and Alternative Applications of Biochar - Chapter 6. Carbons from Biomass for Electrochemical Capacitors - Chapter 7. Carbonaceous Catalysts from Biomass - Chapter 8. Synthesis and Design of Engineered Biochars as Electrode Materials in Energy Storage Systems - Part 4: Biomass Pellets as Fuels - Chapter 9. Biomass Pelletization: Contribution to Renewable Power Generation Scenarios - Chapter 10. Biocarbon Production and Use as A Fuel - Chapter 11. Mechanical Aspects and Applications of Pellets Prepared from Biomass Resources - Part 5: Biosynthesis of Polymers from Renewable Biomass - Chapter 12. Microbial Production and Properties of Lactic Acid-Based Polymers and Oligomers from Renewable Feedstock.
LC Subjects	Biomedical materials. Biomedical engineering. Biotechnology.
Other Subjects	Biomedical Engineering/Biotechnology. Biomaterials. Biotechnology.
Additional formats	Print version: Production of materials from sustainable biomass resources 9789811337673 (DLC) 2019931852 Printed edition: 9789811337673 Printed edition: 9789811337697
Series	Biofuels and Biorefineries, 2214-1537; 9 Biofuels and Biorefineries, 2214-1537; 9

Production of platform chemicals from sustainable resources

LCCN	2019769036
Type of material	Book

Main title	Production of Platform Chemicals from Sustainable Resources / edited by Zhen Fang, Richard L. Smith, Jr., Xinhua Qi.
Edition	1st ed. 2017.
Published/Produced	Singapore: Springer Singapore: Imprint: Springer, 2017.
Description	1 online resource (XVI, 475 pages 139 illustrations, 41 illustrations in color.) PDF
ISBN	9789811041723
Related names	Fang, Zhen, editor. Qi, Xinhua, editor. Smith, Jr., Richard L, editor.
Summary	This book provides state-of-the-art reviews, the latest research, prospects and challenges of the production of platform chemicals such as C6 sugars, 5-hydroxymethylfurfural, furfural, gamma-valerolactone, xylitol, 2,5-furandicarboxylic acid, levulinic acid, ethanol and others from sustainable biomass resources using processes that include heterogeneous catalysis, ionic liquids, hydrothermal/solvothermal, electrochemical and fermentation methods. It also discusses the application of these chemicals and their derivatives for synthesizing commodity chemicals via various routes. Intended as a reference resource for researchers, academicians and industrialists in the area of energy, chemical engineering and biomass conversion, it provides a wealth of information essential for assessing the production and application of various biomass-derived platform chemicals using biological, chemical and electrochemical techniques.
Contents	Part I: Production of Sugars (Chapter 1) - Part II: Production of Aldehydes (Chapters 2-4) - Part III: Production of Acids (Chapters 5-8) - Part IV: Production of Alcohols (Chapters 9-12) - Part V: Production of Lactones and Amino Acids (Chapters 13-14).

LC Subjects	Biochemistry. Forestry. Renewable energy sources.
Other Subjects	Biochemistry, general. Forestry. Renewable and Green Energy.
Additional formats	Print version: Production of platform chemicals from sustainable resources 9789811041716 (DLC) 2017938599 Printed edition: 9789811041716 Printed edition: 9789811041730 Printed edition: 9789811350696
Series	Biofuels and Biorefineries, 2214-1537 Biofuels and Biorefineries, 2214-1537

Sugar: processing, production and uses

LCCN	2018287991
Type of material	Book
Main title	Sugar: processing, production and uses / Fredrick Caras, editor.
Published/Produced	New York: Nova Science Publishers, Inc., [2019]
Description	xiii, 223 pages: illustrations; 23 cm.
ISBN	9781536153965
LC classification	TP377 .S787 2019
Related names	Caras, Fredrick, editor.
Contents	Sugars as a possible indicator of honey authenticity / Dušanka Milojković-Opsenica, Jelena Trifković, Tomislav Tosti and Aleksandra Radoičić - Production of sugars fron sago palm / K.B. Bujang, S.S.A. Adeni and N. Suhaili - The impact of processing on physicochemical characteristics, including metal icons of Brazilian brown sugar / Camila K. de Andrade, Giseli Ducat, Maria L. Felsner, Sueli P. Quináia, Claudio H. Lopes, Maria T.M.R. Borges and Adrilene S. Adriano - Sugarcane production: reuse of by-products for energy purposes / Otávio José de Oliveira, Andréia Marize Rodrigues, Marcelo Girotto Rebelato, Isabel Cristina Rodrigues and Fabrício Pinto Neto

	Filho - Development of sustainable biorefineries of sucrose from sugarcane and beet / Noe Aguilar-Rivera, Ricardo Serna-Lagunes, Maria Elena Galindo-Tovar, Rosalia Nuñez-Pastrana and Luis Alberto Olvera-Vargas - Sugar cane bagasse as a green additive in the production of environmentally friendly elastomeric compounds / Aline Zanchet and Fabiula Danielli Bastos de Sousa.
LC Subjects	Sugar--Manufacture and refining.
Notes	Includes bibliographical references and index.
Series	Food science and technology

Sugarcane: production, properties and uses

LCCN	2020038883
Type of material	Book
Main title	Sugarcane: production, properties and uses / Elisa Russo, editor.
Published/Produced	New York: Nova Science Publishers, 2020.
Description	xii, 200 pages: illustrations; 23 cm
ISBN	9781536184174 (hardcover) (adobe pdf)
LC classification	SB231 .S863 2020
Related names	Russo, Elisa, editor.
Summary	"Sugarcane: Production, Properties and Uses provides details on new industrial technologies of the ethanol extraction process through mathematical modeling for bioenergy crops, socioeconomic and environmental aspects based on Circular Economy concepts, and some commercial and industrial applications in thermal power plants and/or biorefineries. Lignocellulosic biomass is the most abundant of terrestrial feedstocks available for producing chemicals and materials as well as harvesting energy. As such, this compilation aims to describe the structure of sugarcane lignin, as well as its isolation, characterization, and applications. The authors study the co-digestion of sugarcane vinasse and glycerin in a thermophilic anaerobic sequencing

batch biofilm reactor, which has not previously been studied. Additionally, they evaluate the option of using sugarcane cultivars in the production of sucrose in Mexico as a livestock feed source in the dry season, with the determination of variables such as stem height, stem diameter, Brix degrees, health and flowering. Following this, variations of phenol accumulation in healthy sugarcane plants or sugarcane plants experimentally infected with Xanthomonas albilineans, the bacteria that causes leaf scald, are studied. The concluding work evaluates the emissions of gases from a bench diesel engine combustion with different diesel/biodiesel blends from beef tallow/sugarcane diesel in the proportions CD10, CD-B15, CD-B25, CD-B35 and CD-B50, in order to simulate biofuel combustion emissions at different concentrations"-- Provided by publisher.

Contents Perspectives and Challenges of Sugarcane Straw for Bioenergy Production / Julie Brenda Santos da Silva, Harvey Alexander Villa-Vélez, Jorge Bertoldo Júnior, Charles Corrêa Conconi and Glauber Cruz - Lignin from Sugarcane: Extraction, Characterization and Applications / Rafael Garcia Candido - Stability Behavior and Methane Production of a Thermophilic AnSBBR Codigesting Sugarcane Vinasse and Glycerin / J. N. Albuquerque, R. Albanez, G. Lovato, S. M. Ratusznei and J. A. D. Rodrigues - Competitive Management of Sugarcane Varieties for Use in Livestock Feed / Arturo Hernandez-Gastelú and Noé Aguilar-Rivera - Accumulation of Benzoic Acid Derivatives in Sugarcane Tissues as an Active Defense against Xanthomonas albilineans / Carlos Vicente, Dolores Piñón, Roberto de Armas, Elena Sánchez-Elordi, Maritza Martínez and María Estrella Legaz - Gaseous Emissions from the Combustion of Biofuels: Diesel/Beef Tallow Biodiesel/Sugarcane Diesel Blends / Marcelo Silva

	Sthel, Fernanda Gomes Linhares and Marcenilda Amorim Lima.
LC Subjects	Sugarcane--Biotechnology. Biomass energy. Sugarcane as feed.
Notes	Includes bibliographical references and index.
Additional formats	Issued also in electronic format. Online version: Sugarcane Hauppauge, New York: Nova Science Publishers, 2020. 9781536185898 (DLC) 2020038884
Series	Plant science research and practices Plant science research and practices.

Sugarcane: production, properties and uses

LCCN	2020038884
Type of material	Book
Main title	Sugarcane: production, properties and uses / Elisa Russo.
Published/Produced	Hauppauge, New York: Nova Science Publishers, 2020.
Description	1 online resource
ISBN	9781536185898 (adobe pdf) (hardcover)
LC classification	SB231
Related names	Russo, Elisa, editor.
Summary	"Sugarcane: Production, Properties and Uses provides details on new industrial technologies of the ethanol extraction process through mathematical modeling for bioenergy crops, socioeconomic and environmental aspects based on Circular Economy concepts, and some commercial and industrial applications in thermal power plants and/or biorefineries. Lignocellulosic biomass is the most abundant of terrestrial feedstocks available for producing chemicals and materials as well as harvesting energy. As such, this compilation aims to describe the structure of sugarcane lignin, as well as its isolation, characterization, and applications. The authors

study the co-digestion of sugarcane vinasse and glycerin in a thermophilic anaerobic sequencing batch biofilm reactor, which has not previously been studied. Additionally, they evaluate the option of using sugarcane cultivars in the production of sucrose in Mexico as a livestock feed source in the dry season, with the determination of variables such as stem height, stem diameter, Brix degrees, health and flowering. Following this, variations of phenol accumulation in healthy sugarcane plants or sugarcane plants experimentally infected with Xanthomonas albilineans, the bacteria that causes leaf scald, are studied. The concluding work evaluates the emissions of gases from a bench diesel engine combustion with different diesel/biodiesel blends from beef tallow/sugarcane diesel in the proportions CD10, CD-B15, CD-B25, CD-B35 and CD-B50, in order to simulate biofuel combustion emissions at different concentrations"-- Provided by publisher.

Contents

	Combustion of Biofuels: Diesel/Beef Tallow Biodiesel/Sugarcane Diesel Blends / Marcelo Silva Sthel, Fernanda Gomes Linhares and Marcenilda Amorim Lima.
LC Subjects	Sugarcane--Biotechnology.
	Biomass energy.
	Sugarcane as feed.
Notes	Includes bibliographical references and index.
Additional formats	Issued also in electronic format.
	Print version: Sugarcane Hauppauge, New York: Nova Science Publishers, 2020. 9781536184174 (DLC) 2020038883
Series	Plant science research and practices
	Plant science research and practices.

Sustainability assessment of renewables-based products: methods and case studies

LCCN	2015049970
Type of material	Book
Uniform title	Renewables-based technology
Main title	Sustainability assessment of renewables-based products: methods and case studies / edited by Jo Dewulf, Steven De Meester, Rodrigo Alvarenga.
Published/Produced	Chichester, UK; Hoboken, NJ: John Wiley & Sons, 2016.
Description	1 online resource.
ISBN	9781118933923 (ePub)
	9781118933930 (Adobe PDF)
LC classification	TJ808
Related names	Dewulf, Jo, editor.
	Meester, Steven De, 1985- editor.
	Alvarenga, Rodrigo, editor.
Contents	The growing role of biomass for future resource supply: prospects and pitfalls - Helmut Haber - The growing role of photovoltaic solar, wind and geothermal energy as renewables for electricity generation / W.G.J.H.M. van Sark, J.G. Schepers, J.D.A.M. van Wees - Assessment of sustainability within holistic process design / Alexei Lapkin,

Philipp-Maximilian Jacob, Polina Yaseneva, Charles Gordon, Amy Peace - A mass balance approach to link sustainable renewable resources in chemical synthesis with market demand / Dr. Claudius Kormann - Early R&D stage sustainability assessment: the 5-pillar method / Akshay D. Patel, John A. Posada, Li Shen, Martin K. Patel - Assessing the sustainability of land use: a systems approach / Miguel Brando - Water use analysis / Francesca Verones, Stephan Pfister, Markus Berger - Material intensity of food production and consumption / Lucia Mancini and Michael Lettenmeier - Material and energy flow analysis / Goto, Naohiro, Nova Ulhasanah, Hirotsugu Kamahara, Udin Hasanudin, Ryuichi Tachibana, Koichi Fujie - Exergy and cumulative exergy use analysis / Sofie Huysman, Thomas Schaubroeck, Jo Dewulf - Carbon and environmental footprint methods for renewables-based products and transition pathways to 2050 / Geoffrey P. Hammond - Tracking supply and demand of biocapacity through ecological footprint accounting / David Lin, Alessandro Galli, Michael Borucke, Elias Lazarus, Nicole Grunewald, Jon Martindill, David Zimmerman, Serena Mancini, Katsunori Iha, and Mathis Wackernagel - Life cycle assessment and sustainability: supporting decision making by business and policy / Sala Serenella, Fabrice Mathieux, Rana Pant - Life cycle costing / Andreas Ciroth, Jutta Hildenbrand, Bengt Steen - Social life cycle assessment: methodologies and practice / Alessandra Zamagni, Pauline Feschet, Anna Irene De Luca, Nathalie Iofrida, Patrizia Butto - Life cycle assessment of solar technologies / F. Ardente, M. Cellura, S. Longo, M. Mistretta - Asssessing the sustainability of geothermal utilization / Ruth Shortalla, Gudni Axelsson and Brynhildur Davidsdottira - Biofuels from terrestrial biomass: sustainability assessment

	of sugarcane biorefineries in Brazil / Otavio Cavalett, Marcos D.B. Watanabe, Alexandre Souza, Mateus F. Chagas, Tassia L. Junqueira, Antonio Bonomi - Algae as promising biofeedstock: searching for sustainable production processes and market applications / Sue Ellen Taelman, Steven De Meester, Jo Dewulf - Life cycle assessment of biobased and fossil-based succinic acid / Marieke Smidt, Jeroen den Hollander, Henk Bosch, Yang Xiang, Maarten van der Graaf, Anne Lambin, Jean-Pierre Duda - Biobased poly vinylchloride (PVC) / Rodrigo A. F. Alvarenga, Zdenek Hruska, Alain Wathelet, Jo Dewulf - Evaluation of wood cascading / Karin Heglmeier, Gabriele Weber-Blaschke, Klaus Richter - Time-dependent life-cycle assessment of bio-based packaging materials / Maartje N. Sevenster.
LC Subjects	Renewable energy sources. Green chemistry. Green products. Sustainable engineering.
Notes	"An extensive update and sequel to the successful 2006 book: Renewables-Based Technology: Sustainability Assessment." Includes bibliographical references and index.
Additional formats	Print version: Chichester, UK; Hoboken, NJ: John Wiley & Sons, 2016 9781118933947 (DLC) 2015035081

Sustainability assessment of renewables-based products: methods and case studies

LCCN	2015035081
Type of material	Book
Uniform title	Renewables-based technology
Main title	Sustainability assessment of renewables-based products: methods and case studies / edited by Jo Dewulf, Sustainability Assessment Unit, Institute for Environment and Sustainability, Joint Research

Centre, European Commission, Italy and Research Group Envoc, Ghent University, Belgium, Steven De Meester, Research Group Envoc, Ghent University, Belgium, Rodrigo A.F. Alvarenga, Departamento de Engenharia Ambiental, Universidade do Estado de Santa, Catarina (UDESC), Brazil and Departamento de Engenharia Sanitária e Ambiental, Universidade Federal de Santa Catarina (UFSC), Brazil, and EnCiclo Soluções Sustentáveis, Brazil.

Published/Produced Chichester, UK; Hoboken, NJ: Wiley, 2016.

Description xxvi, 368 pages; 25 cm

ISBN 9781118933947 (cloth)

LC classification TJ808 .R47 2016

Related names Dewulf, Jo, editor.
Meester, Steven De, 1985- editor.
Alvarenga, Rodrigo, editor.

Contents The growing role of biomass for future resource supply: prospects and pitfalls - Helmut Haber - The growing role of photovoltaic solar, wind and geothermal energy as renewables for electricity generation / W.G.J.H.M. van Sark, J.G. Schepers, J.D.A.M. van Wees - Assessment of sustainability within holistic process design / Alexei Lapkin, Philipp-Maximilian Jacob, Polina Yaseneva, Charles Gordon, Amy Peace - A mass balance approach to link sustainable renewable resources in chemical synthesis with market demand / Dr. Claudius Kormann - Early R&D stage sustainability assessment: the 5-pillar method / Akshay D. Patel, John A. Posada, Li Shen, Martin K. Patel - Assessing the sustainability of land use: a systems approach / Miguel Brando - Water use analysis / Francesca Verones, Stephan Pfister, Markus Berger - Material intensity of food production and consumption / Lucia Mancini and Michael Lettenmeier - Material and energy flow analysis / Goto, Naohiro, Nova Ulhasanah, Hirotsugu Kamahara, Udin Hasanudin, Ryuichi

Tachibana, Koichi Fujie - Exergy and cumulative exergy use analysis / Sofie Huysman, Thomas Schaubroeck, Jo Dewulf - Carbon and environmental footprint methods for renewables-based products and transition pathways to 2050 / Geoffrey P. Hammond - Tracking supply and demand of biocapacity through ecological footprint accounting / David Lin, Alessandro Galli, Michael Borucke, Elias Lazarus, Nicole Grunewald, Jon Martindill, David Zimmerman, Serena Mancini, Katsunori Iha, and Mathis Wackernagel - Life cycle assessment and sustainability: supporting decision making by business and policy / Sala Serenella, Fabrice Mathieux, Rana Pant - Life cycle costing / Andreas Ciroth, Jutta Hildenbrand, Bengt Steen - Social life cycle assessment: methodologies and practice / Alessandra Zamagni, Pauline Feschet, Anna Irene De Luca, Nathalie Iofrida, Patrizia Butto - Life cycle assessment of solar technologies / F. Ardente, M. Cellura, S. Longo, M. Mistretta - Asssessing the sustainability of geothermal utilization / Ruth Shortalla, Gudni Axelsson and Brynhildur Davidsdottira - Biofuels from terrestrial biomass: sustainability assessment of sugarcane biorefineries in Brazil / Otavio Cavalett, Marcos D.B. Watanabe, Alexandre Souza, Mateus F. Chagas, Tassia L. Junqueira, Antonio Bonomi - Algae as promising biofeedstock: searching for sustainable production processes and market applications / Sue Ellen Taelman, Steven De Meester, Jo Dewulf - Life cycle assessment of biobased and fossil-based succinic acid / Marieke Smidt, Jeroen den Hollander, Henk Bosch, Yang Xiang, Maarten van der Graaf, Anne Lambin, Jean-Pierre Duda - Biobased poly vinylchloride (PVC) / Rodrigo A. F. Alvarenga, Zdenek Hruska, Alain Wathelet, Jo Dewulf - Evaluation of wood cascading / Karin Heglmeier, Gabriele Weber-Blaschke, Klaus

Richter - Time-dependent life-cycle assessment of bio-based packaging materials / Maartje N. Sevenster.

LC Subjects Renewable energy sources.
Green chemistry.
Green products.
Sustainable engineering.

Notes "An extensive update and sequel to the successful 2006 book: Renewables-Based Technology: Sustainability Assessment."
Includes bibliographical references and index.

Additional formats Online version: Chichester, UK; Hoboken, NJ: John Wiley & Sons, 2016 9781118933923 (DLC) 2015049970

Sustainable biotechnology- enzymatic resources of renewable energy

LCCN 2019742321

Type of material Book

Main title Sustainable Biotechnology- Enzymatic Resources of Renewable Energy / edited by Om V. Singh, Anuj K. Chandel.

Edition 1st ed. 2018.

Published/Produced Cham: Springer International Publishing: Imprint: Springer, 2018.

Description 1 online resource (XII, 476 pages 65 illustrations, 45 illustrations in color.)
PDF

ISBN 9783319954806

Related names Chandel, Anuj K. editor.
Singh, Om V. editor.

Summary Nature offers abundant renewable resources that can be used to partially replace fossil fuels and commodity chemicals but issues of cost, technology readiness levels, and compatibility with existing distribution networks remain huge challenges. Cellulosic ethanol and biodiesel are the most immediately obvious target fuels, with hydrogen, methane and butanol as other potentially viable products. This book continues to bridge the technology gap and focus on critical aspects of lignocellulosic biomolecules and the respective mechanisms regulating their bioconversion to liquid fuels into energy and value-added products of industrial significance. This book is a collection of reviews elucidating several broad-ranging areas of progress and challenges in the utilization of sustainable resources of renewable energy, especially in biofuels. This book comes just at a time when government and industries are accelerating their efforts in the exploration of alternative energy resources, with expectations of the establishment of long-term sustainable alternatives to petroleum-based liquid fuels. Apart from liquid fuel this book also emphasizes the use of sustainable resources for value-added products, which may help in revitalizing the biotechnology industry at a broader scale. This book also provides a comprehensive review of basic literature and advance research methodologies to graduate students studying environmental microbiology, chemical engineering, bio-economy and microbial biotechnology.

Contents

1. Introduction - 2. Role of Systematic Biology in Biorefining of Lignocellulosic Residues for Biofuels and Chemicals Production - 3. Biotechnological advancements in cellulosic ethanol production - 4. Sustainable production of biofuels from weedy biomass and other unconventional lignocellulose wastes - 5. Technological Aspects of lignocellulose conversion into biofuels: Key challenges and practical solutions - 6. Beyond ethanol: Contribution of various bioproducts to enhance the viability of biorefineries - 7. Real Techno-economic assessment of bioethanol production from major lignocellulosic residues under different process configurations - 8. Lignocellulolytic Enzymes from Thermophiles - 9. Application of enzymes in Sustainable liquid transportation fuels

	production - 10. The Realm of Lipases in Biodiesel Production - 11. Nanotechnology- based developments in biofuel production: Current trends and applications - 12. Ester-based biofuels from wastes - 13. Sustainable production of biogas from renewable sources: global overview, scale up opportunities and potential market trends - 14. Microbially Originated Polyhydroxyalkanoate (PHA) Biopolymers: An Insight into the Molecular Mechanism and Biogenesis of PHA Granules - 15. Biopolymer Synthesis and Biodegradation - 16. The application of microbial consortia in a biorefinery context: understanding the importance of artificial lichens - 17. Green microalgae as substrate for producing biofuels and chlorophyll in biorefineries - 18. Future perspectives and potential applications of enzymes in sericulture - index.
LC Subjects	Biotechnology. Microbiology. Renewable energy sources.
Other Subjects	Microbiology. Biotechnology. Renewable and Green Energy.
Additional formats	Print version: Sustainable biotechnology-enzymatic resources of renewable energy. 9783319954790 (DLC) 2018948599 Printed edition: 9783030070434 Printed edition: 9783319954790 Printed edition: 9783319954813

Sustainable catalysis for biorefineries

LCCN	2018304176
Type of material	Book
Main title	Sustainable catalysis for biorefineries / edited by Francesco Frusteri, Donato Aranda and Giuseppe Bonura.
Published/Produced	London: Royal Society of Chemistry, [2018]
Description	xiv, 372 pages: illustrations; 25 cm.
ISBN	1782629637 hardcover

	9781782629634 hardcover
LC classification	TP155.2.E58 S87 2018
Related names	Frusteri, Francesco, editor.
	Aranda, Donato, editor.
	Bonura, Giuseppe, editor.
Summary	This book explores the most effective or promising catalytic processes for the conversion of biobased components into high added value products, as platform chemicals and intermediates.
LC Subjects	Green chemistry.
	Chemical processes.
	Catalysis.
Other Subjects	Catalysis.
	Chemical processes.
	Green chemistry.
Notes	Includes bibliographical references and index.
Additional formats	ebook version: 9781788013567
Series	Green chemistry series; no. 56
	RSC green chemistry series; 56.

Sustainable energy technology and policies: a transformational journey, volume 1

LCCN	2019747075
Type of material	Book
Main title	Sustainable Energy Technology and Policies: A Transformational Journey, Volume 1 / edited by Sudipta De, Santanu Bandyopadhyay, Mohsen Assadi, Deb A Mukherjee.
Edition	1st ed. 2018.
Published/Produced	Singapore: Springer Singapore: Imprint: Springer, 2018.
Description	1 online resource (XX, 450 pages 178 illustrations, 110 illustrations in color.)
	PDF
ISBN	9789811071881
Related names	Assadi, Mohsen. editor.
	Bandyopadhyay, Santanu. editor.
	De, Sudipta. editor.
	Mukherjee, Deb A. editor.

Summary

This book presents a state-of-the-art compilation focusing on both technological and policy aspects of sustainable energy production and consumption, which deals with issues like the need for and planning of smart cities, alternative transport fuel options, sustainable power production, pollution control technologies et cetera The book comprises contributions from experts from all over the world, and addresses energy sustainability from different viewpoints. Specifically, the book focuses on energy sustainability in the Indian scenario with a background of the global perspective. Contributions from academia, policy makers and industry are included to address the challenge from different perspectives. The contents of this book will prove useful to researchers, professionals, and policy makers working in the area of green and sustainable energy.

Contents

	Review on Upgrading of Lignin Derived Bio-Oil Model Components - 15. Sustainability of microalgal bio-refinery: Scope, challenges and opportunities - 16. Indian Agro-wastes for 2G-Biorefineries: Strategic Decision on Conversion Processes - 17. Utilization of light pipe in Indian buildings - sustainability through passive system - 18. Thermal performance evaluation of building roofs embedded PCM for multi-climatic zones - 19. Energy Efficiency at India's Largest Milk Processing Plant.
LC Subjects	Energy and state. Energy policy. Renewable energy sources. Transportation.
Other Subjects	Renewable and Green Energy. Energy Policy, Economics and Management. Transportation.
Additional formats	Print version: Sustainable energy technology and policies. 9789811071874 (DLC) 2017958612 Printed edition: 9789811071874 Printed edition: 9789811071898 Printed edition: 9789811356018
Series	Green Energy and Technology, 1865-3529 Green Energy and Technology, 1865-3529

The chemistry of bio-based polymers

LCCN	2019051488
Type of material	Book
Personal name	Fink, Johannes Karl, author.
Main title	The chemistry of bio-based polymers / Johannes Karl Fink, Montanuniversität Leoben, Austria.
Edition	Second edition.
Published/Produced	Hoboken, New Jersey: John Wiley & Sons, Inc.; Salem, Massachusetts: Scrivener Publishing LLC, [2020]
ISBN	9781119681311 (cloth) (adobe pdf) (epub)

LC classification	TP1180.B55 F56 2020
Summary	"The recent explosion of interdisciplinary research has fragmented the knowledge base surrounding renewable polymers. The Chemistry of Bio-based Polymers 2nd edition brings together, in one volume, the research and work of Professor Johannes Fink, focusing on biopolymers that can be synthesized from renewable polymers. After introducing general aspects of the field, the book's subsequent chapters examine the chemistry of biodegradable polymeric types sorted by their chemical compounds, including the synthesis of low molecular compounds. Various categories of biopolymers are detailed including vinyl-based polymers, acid and lactone polymers, ester and amide polymers, carbohydrate-related polymers and others. Procedures for the preparation of biopolymers and biodegradable nanocomposites are arranged by chemical methods and in vitro biological methods, with discussion of the issue of "plastics from bacteria." The factors influencing the degradation and biodegradation of polymers used in food packaging, exposed to various environments, are detailed at length. The book covers the medical applications of bio-based polymers, concentrating on controlled drug delivery, temporary prostheses, and scaffolds for tissue engineering. Professor Fink also addresses renewable resources for fabricating biofuels and argues for localized biorefineries, as biomass feedstocks are more efficiently handled locally"--Provided by publisher.
LC Subjects	Biodegradable plastics. Biopolymers.
Notes	Includes bibliographical references and index.
Additional formats	Online version: Fink, Johannes Karl, The chemistry of bio-based polymers Second edition. Hoboken, New Jersey: John Wiley & Sons, Inc., [2020] 9781119681298 (DLC) 2019051489

The chemistry of bio-based polymers

LCCN	2019051489
Type of material	Book
Personal name	Fink, Johannes Karl, author.
Main title	The chemistry of bio-based polymers / Johannes Karl Fink, Montanuniversität Leoben, Austria.
Edition	Second edition.
Published/Produced	Hoboken, New Jersey: John Wiley & Sons, Inc.; Salem, Massachusetts: Scrivener Publishing LLC, [2020]
Description	1 online resource
ISBN	9781119681267 (epub) 9781119681298 (adobe pdf) (cloth)
LC classification	TP1180.B55
Summary	"The recent explosion of interdisciplinary research has fragmented the knowledge base surrounding renewable polymers. The Chemistry of Bio-based Polymers 2nd edition brings together, in one volume, the research and work of Professor Johannes Fink, focusing on biopolymers that can be synthesized from renewable polymers. After introducing general aspects of the field, the book's subsequent chapters examine the chemistry of biodegradable polymeric types sorted by their chemical compounds, including the synthesis of low molecular compounds. Various categories of biopolymers are detailed including vinyl-based polymers, acid and lactone polymers, ester and amide polymers, carbohydrate-related polymers and others. Procedures for the preparation of biopolymers and biodegradable nanocomposites are arranged by chemical methods and in vitro biological methods, with discussion of the issue of "plastics from bacteria." The factors influencing the degradation and biodegradation of polymers used in food packaging, exposed to various environments, are detailed at length. The book covers the medical applications of bio-based

	polymers, concentrating on controlled drug delivery, temporary prostheses, and scaffolds for tissue engineering. Professor Fink also addresses renewable resources for fabricating biofuels and argues for localized biorefineries, as biomass feedstocks are more efficiently handled locally"-- Provided by publisher.
LC Subjects	Biodegradable plastics. Biopolymers.
Notes	Includes bibliographical references and index.
Additional formats	Print version: Fink, Johannes Karl. The chemistry of bio-based polymers Second edition. Hoboken, New Jersey: John Wiley & Sons, Inc.; Salem, Massachusetts: Scrivener Publishing LLC, [2020] 9781119681311 (DLC) 2019051488

Virtual biorefinery: an optimization strategy for renewable carbon valorization

LCCN	2019756761
Type of material	Book
Main title	Virtual Biorefinery: An Optimization Strategy for Renewable Carbon Valorization / edited by Antonio Bonomi, Otavio Cavalett, Marcelo Pereira da Cunha, Marco A. P. Lima.
Edition	1st ed. 2016.
Published/Produced	Cham: Springer International Publishing: Imprint: Springer, 2016.
Description	1 online resource (XL, 285 pages 83 illustrations) PDF
ISBN	9783319260457
Related names	Bonomi, Antonio. editor. Cavalett, Otavio. editor. Lima, Marco A. P. editor. Pereira da Cunha, Marcelo. editor.
Summary	This book presents a concise framework for assessing technical and sustainability impacts of existing biorefineries and provides a possible road map for development of novel biorefineries. It offers a detailed, integrated approach to evaluate

	the entire biomass production chain, from the agricultural feedstock production and transportation, to the industrial conversion and commercialization and use of products. The Brazilian sugarcane biorefinery is used as a case study; however, the methods and concepts can be applied to almost any biomass alternative. Chapters explore the main issues regarding biorefinery assessment, including feedstock production and transportation modeling, biofuels and green chemistry products, as well as assessment of sustainability impacts. This book is a valuable source of information to researchers in bioenergy, green chemistry and sustainability fields. It also provides a useful framework for government agencies, investors and the energy industry to evaluate and predict the success of current and future biorefinery alternatives.
Contents	Background - The Virtual Sugarcane Biorefinery Concept - The Agricultural Production Model - Biorefinery Alternatives - Biorefinery Products Logistics, Commercialization and Use - Sustainability Assessment Methodologies - Use of the VSB to Assess Biorefinery Strategies - Use of VSB to Plan Research Programs and Public Policies - Final Remarks.
LC Subjects	Fossil fuels. Renewable energy sources.
Other Subjects	Fossil Fuels (incl. Carbon Capture). Renewable and Green Energy. Renewable and Green Energy.
Additional formats	Print version: Virtual biorefinery. 9783319260433 (DLC) 2015953802 Printed edition: 9783319260433 Printed edition: 9783319260440 Printed edition: 9783319798813
Series	Green Energy and Technology, 1865-3529 Green Energy and Technology, 1865-3529

Waste biorefinery: integrating biorefineries for waste valorisation

LCCN	2020931490
Type of material	Book
Main title	Waste biorefinery: integrating biorefineries for waste valorisation / Thallada Bhaskar, Ashok Pandey, Eldon R Rene, Daniel Tsang.
Edition	1.
Published/Produced	Waltham: Elsevier, 2020.
ISBN	9780128182284 (paperback) (epub)
Related names	Bhaskar, Thallada, editor. Pandey, Ashok, editor. Rene, Eldon R, editor. Tsang, Daniel, editor.
Summary	"Waste Biorefinery: Integrating Biorefineries for Waste Valorisation provides the various options available for several renewable waste streams. The book includes scientific and technical information pertaining to the most advanced and innovative processing technologies used for the conversion of biogenic waste to biofuels, energy products and biochemicals. In addition, the book reports on recent developments and new achievements in the field of biochemical and thermo-chemical methods and the necessities and potential generated by different kinds of biomass in presumably more decentralized biorefineries. The book presents an assortment of case-studies from developing and developed countries pertaining to the use of sustainable technologies for energy recovery from different waste matrices. Advantages and limitations of different technologies are also discussed by considering the local energy demands, government policies, environmental impacts, and education in bioenergy"-- Provided by publisher.

Waste biorefinery: potential and perspectives

LCCN	2020304668
Type of material	Book

Main title	Waste biorefinery: potential and perspectives / [edited by] Thallada Bhaskar, Ashok Pandey, S. Venkata Mohan, Duu-Jong Lee, Samir Kumar Khanal.
Published/Produced	Amsterdam, Netherlands; Cambridge, MA: Elsevier, [2018]
Description	xx, 870 pages: illustrations; 24 cm
ISBN	9780444639929 (paperback) (ePub ebook)
LC classification	TP339 .W375 2018
Related names	Bhaskar, Thallada, editor. Pandey, Ashok, editor. Mohan, S. Venkata, editor. Lee, D. J. (Duu-Jong), editor. Khanal, Samir Kumar, editor.
Contents	Dedicated and Waste Feedstocks for Biorefinery: An Approach to Develop a Sustainable Society / Kirtikumar C. Badgujar, Bhalchandra M. Bhanage - Kinetic Analysis of Biomass Pyrolysis / Vaibhav Dhyani, Thallada Bhaskar - Sidestreams From Bioenergy and Biorefinery Complexes as a Resource for Circular Bioeconomy / Lakhya J. Konwar, Jyri-Pekka Mikkola, Neonjyoti Bordoloi, Ruprekha Saikia, Rahul S. Chutia, Rupam Kataki - Thermochemical Conversion Processes for Waste Biorefinery / Kawnish Kirtania - Combined Gasification-Fermentation Process in Waste Biorefinery / Konstantinos Chandolias, Tobias Richards, Mohammad J. Taherzadeh - Acidogenic Biorefinery: Food Waste Valorization to Biogas and Platform Chemicals / Omprakash Sarkar, Sai K. Butti, S. Venkata Mohan - Food Supply Chain Waste: A Functional Periodic Table of Bio-Based Resources / Avtar S. Matharu, Eduardo M. de Melo, Joseph A. Houghton - Conversion of Solid Wastes to Fuels and Chemicals Through Pyrolysis / Sushil Adhikari, Hyungseok Nam, Jyoti P. Chakraborty - Lignocellulosic Biorefinery Wastes, or Resources? / Anil K. Mathew, Amith Abraham,

Kiran Kumar Mallapureddy, Rajeev K. Sukumaran - Process Intensification and Process Integration for Hydrothermal Processing of Forest Residues and Agricultural Wastes / Khanh-Quang Tran - Pyrolysis of Lignocellulosic Biomass for Biochemical Production / Joo-Sik Kim, Gyung-Goo Choi - Copyrolysis of Lignocellulosic Biomass With Waste Plastics for Resource Recovery / Deepak K. Ojha, Ravikrishnan Vinu - Microwave-Driven Biorefinery for Utilization of Food and Agricultural Waste Biomass / Shuntaro Tsubaki, Jun-ichi Azuma, Satoshi Fujii, Rawel Singh, Bhaskar Thallada, Yuji Wada - Lignin Conversion: A Key to the Concept of Lignocellulosic Biomass-Based Integrated Biorefinery / Ejaz Ahmad, Kamal K. Pant - Synthesis of Bioethanol From Invasive Weeds: Process Design, Optimization, and Intensification With Ultrasound / Shuchi Singh, Arun Goyal, Vijayanand S. Moholkar - Valorization of Wastes From Agrofood and Pulp and Paper Industries Within the Biorefinery Concept: Southwestern Europe Scenario / Aloia Romaní, Michele Michelin, Lucília Domingues, José A. Teixeira - Ionic Liquid Stable Cellulases and Hemicellulases: Application in Biobased Production of Biofuels / Ayesha Sadaf, Jasneet Grewal, Sunil K. Khare - Resource Recovery From Wastes and Wastewaters Using Bioelectrochemical Systems / Jai Sankar Seelam, Sausan A. Maesara, Gunda Mohanakrishna, Sunil A. Patil, Annemiek ter Heijne, Deepak Pant - Exploitation and Biorefinery of Microalgae / Revathy Sankaran, Pau L. Show, Dillirani Nagarajan, Jo-Shu Chang - Waste Biorefinery in Arid/Semiarid Regions / Juan-Rodrigo Bastidas-Oyanedel, Jens Ejbye Schmidt - Potential and Perspective of Castor Biorefinery / Satya N. Naik, Devesh K. Saxena, Bhushan R. Dole, Sunil K. Khare - Insect-Based Biorefinery for

	Bioenergy and Bio-Based Products / Karthik Rajendran, K.C. Surendra, Jeffery K. Tomberlin, Samir K. Khanal - Thermochemical Valorization of Paper Deinking Residue through Microwave-Assisted Pyrolysis / Zhanrong Zhang, Avtar S. Matharu - Process Design and Integration Philosophy for Competitive Waste Biorefineries: Example of Levulinic Acid Production From Representative Lignocellulosic Biomasses / Elias Martinez-Hernandez, Jhuma Sadhukhan - Monitoring and Control of Bioethanol Production From Lignocellulosic Biomass / James Gomes, Jyoti Batra, Viki R. Chopda, P. Kathiresan, Anurag S. Rathore - Environmental Impacts of Utilizing Woody Biomass for Energy: A Case Study in Japan / Tomohiro Tabata.
LC Subjects	Biomass energy.
Notes	Includes bibliographical references and indexes.

Waste biorefinery: value addition through resources utilization

LCCN	2021936779
Type of material	Book
Main title	Waste biorefinery: value addition through resources utilization / Thallada Bhaskar, Sunita Varjani, Ashok Pandey, Eldon R. Rene.
Edition	1.
Published/Produced	Cambridge: Elsevier, 2021.
ISBN	9780128218792 (paperback) (ebook)
Related names	Bhaskar, Thallada, editor. Varjani, Sunita, editor. Pandey, Ashok, editor. Rene, Eldon R., editor.
Summary	"Waste Biorefinery: Value Addition through Resources Utilization provides scientific and technical information surrounding the most advanced and innovative processing technologies used for the conversion of biogenic waste to biofuels, energy products and biochemicals. The

book covers recent developments and achievements in the field of biochemical, thermo-chemical and hybrid methods and the necessities and potentials generated by different kinds of residual streams, including biomass in presumably more decentralized biorefineries. An assortment of case-studies from developing and developed countries illustrate the topics presented, covering energy, chemicals, fuels, food for animal recovery from different waste matrices, and more. Finally, the advantages and limitations of different technologies are discussed, considering local energy demand, government policies, environmental impacts and education in bioenergy. This book will serve as an excellent resource for science graduates, chemical engineers, environmental engineers, biotechnologists and industrial experts in these areas"-- Provided by publisher.

Index

#

A

B

C

M

N

O

P

R

S

T

V

W